Laboratory Manual

Robotics

Theory and Industrial Applications

Second Edition

Larry (Tim) Ross
Department Chair—Department of Technology
Eastern Kentucky University
Richmond, KY

Stephen W. Fardo
Foundation Professor—Department of Technology
Eastern Kentucky University
Richmond, KY

Sergio D. Sgro
Assistant Professor—Department of Technology
Eastern Kentucky University
Richmond, KY

James W. Masterson
Professor Emeritus—Department of Technology
Eastern Kentucky University
Richmond, KY

Robert L. Towers
Retired Professor—Department of Technology
Eastern Kentucky University
Richmond, KY

Publisher
The Goodheart-Willcox Company, Inc.
Tinley Park, Illinois
www.g-w.com

Manufactured in the United States of America.

ISBN 978-1-60525-322-0

3 4 5 6 7 8 9 – 11 – 15 14 13

Introduction

This Laboratory Manual is a supplement for the text ***Robotics: Theory and Industrial Applications***. It is intended to help you carry out an in-depth study of robotic systems and the subsystems which comprise them.

Robotics is a very comprehensive, applications-oriented field of study. A complete understanding of this field involves many different technical areas such as electrical principles, electronic devices, digital principles, electromechanical fundamentals, basic programming techniques, hydraulics, pneumatics, and basic manufacturing processes.

The laboratory activities in this manual will help you better understand many of these topics. The activities in Unit I—Principles of Robotics, describe robotic programming. The laboratory activities in Units II, III, and IV are hands-on activities dealing with related technical principles of robotic sub-systems.

The software discussion will provide you with an opportunity to learn to compile simple robot programs, as well as simulate the actions of a robot. You will be working as if you were writing a program away from a robot's workstation (a common occurrence in industry). A robot is not required but would be desirable for further study.

Each laboratory activity is organized as follows:

- Objective. Outlines what the students are expected to learn when the experiment is completed.
- Procedure. Provides logical step-by-step sequence to complete the learning activity. Charts and tables are provided to aid in data recording.
- Analysis. Provides specific questions and problems that supplement the experimental activity.

This manual and ***Robotics: Theory and Industrial Applications*** textbook should provide a framework for your training in robotics. You should supplement your knowledge by additional reading about the technical basics of robotics. We hope you will find this Laboratory Manual a valuable tool in your learning experiences as a student of robotics.

Larry (Tim) Ross
Stephen W. Fardo
Sergio D. Sgro
James W. Masterson
Robert L. Towers

Table of Contents

Unit I—Principles of Robotics

Unit II—Power Supplies and Movement Systems

Activity 3-1—Programming Environment

Name ______________________ Date ______________ Score ____________

Objectives:

The purpose of this exercise is to provide information about working with a typical programming language. The knowledge you gain from studying this programming language will help you have a better understanding of other robot programming languages.

After completing this exercise, you will be able to:

- Identify the programming environment, and move among program menus.
- Create a robot program, teach point location positions, successfully compile a robot program, and run it on a program simulator.

Procedure:

Note: The software being used for this lab manual is the MELFA BASIC IV software. The procedure that follows is used to show programming methods.

1. Turn on the computer and monitor.
2. Select the programming icon from the desktop.
3. The **RT Toolbox** dialog box will appear on screen, **Figure 3-1**.

Figure 3-1. The **RT Toolbox** dialog box.

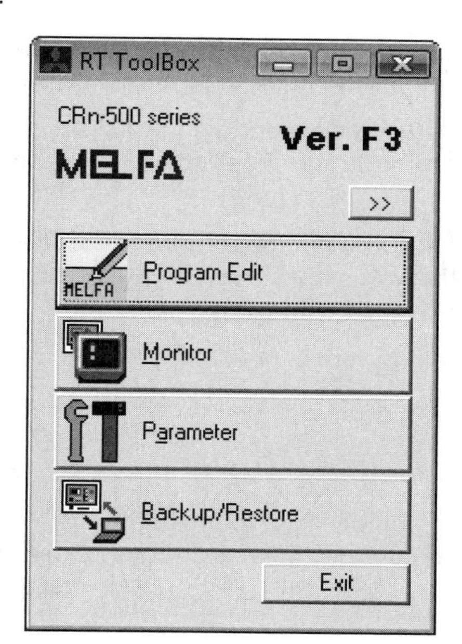

4. Find and select **Program Edit**.
5. Once the **Program Edit** screen is open, find and select the **Help Menu**.
6. The **Help Menu** often contains a wealth of information about programming keywords and commands and examples of how these functions are used in a typical program.

   When you are familiar with the **Help Menu**, you can start learning about programming. We will begin by creating a short program and saving it. We will then discuss how to teach points, compile a program, and run it as a simulation.

7. Exit the **Help Menu** and return to the **Program Edit** screen. Make sure you are in **BASIC** mode by going to the **Window** menu and selecting **BASIC editor**. MELFA BASIC IV software allows uppercase letters and lowercase letters. You must push the [Enter] key after each line of code, or the code will not be saved in the program. You will find that each software will have its own syntax and other specific programming characteristics. The following is an example of a program for the MELFA BASIC IV software. This program will move a part from one location to another.

```
10 MOV PSAFE
20 MOV P1
30 MVS P1, -30
40 DLY .5
50 HCLOSE 1
60 DLY .5
70 MOV P1
80 MOV P2
90 MVS P2, -30
100 DLY .5
110 HOPEN 1
120 DLY .5
130 MVS P2
140 MOV PSAFE
150 END
```

8. After developing a program using the software for your robot application, be sure to save the program. To save the file, select **File**, **Save**, then type in the file name AC01.prg (be sure to use the extension required by your robot application). A dialog box will indicate whether any syntax errors were found in the program. The message No syntax error was found indicates the program was entered correctly. If the software finds an error in your program, an error message with a line number will be displayed. Promptly correct any issues before proceeding and save the file.
9. The next step is to teach the points that will be used by the program to move the part from location P1 to location P2. Again, each programming software will perform this step in a different manner.
10. If your computer is connected to the robot controller, you can download the position directly to the controller. In this example, you will write the program as if you were writing a program away from the robot's work cell.
11. If your robot software has a simulator, you will be able to see the robot moving to the location of the points entered. The simulator provides the user with a visual representation of how the robot moves within the work envelope. The work envelope is the working area of the robot where points can be taught.
12. After a successful compilation, the program is now ready to run in the simulator or to be downloaded to the robot.

Analysis:

For the robotics software you are using, complete the following.

1. What programming statement (command) causes the robot to move to a point location?

____

2. When teaching points, what methods can be used to move the robot to a particular point location?

____

____

3. If you are having trouble with the use of a command, what menu item would you use to find out how to use the command properly?

____

4. What is a work envelope?

____

Activity 3-2—Axis Movement Commands

Name ______________________________ Date ______________ Score ___________

Objectives:

The purpose of this exercise is to provide information about basic robotic movement programming that allows for one of the most common industrial applications: pick and place.
After completing this exercise, you will be able to:

- Write a program that moves the robot from one location to another with both curvilinear and straight movements.
- Incorporate a retraction command in the tool direction (Z-axis) to identify a point not specifically taught to the robot.
- Open and close the robot's end-effector.

Programming Syntax:

MOV	Move
MVS	Move straight
HOPEN	Open the robot's end-effector
HCLOSE	Close the robot's end-effector
HLT	Paused execution of program, program restarts on next line
END	Defines end of a program

Examples of Programming Syntax:

MOV P3	Robot moves to point called P3 (robot "decides" best path) taking shortest path; known as joint interpolation
MOV P4, -30	Robot moves from position P4 to a retracted 30 mm in the tool direction (Z-axis); known as linear interpolation
MVS P3	Robot moves straight to P3 from its current location
MVS, -25	Moves from the current position to a position retracted 25 mm in the tool direction (straight up)
HOPEN 1	Opens the motorized hand control 1
HOPEN 1, 63, 63, .5	Opens the pneumatic hand 1 with 3.5 kg (starting grasp force), 3.5 kg (holding grasp force), and .5 seconds (grasp force holding time)
HCLOSE 1	Closes the motorized hand control 1

Procedure:

Note: The software being used for this lab manual is the MELFA BASIC IV software. The procedure that follows is used to show programming methods.

Write a program that moves a block from location A (P1) to location B (P2). When the block is placed, the robot returns to the safe position (**PSAFE**). The robot must then run in reverse, pick up the block at location B, and place it at its original location A. Use the following commands at least once in your program:

- **MOV**
- **MVS**
- **HOPEN**
- **HCLOSE**
- **END**

Use the example below to begin your program:

```
10 'PSAFE is the safe position out of the way of obstructions
20 'P1 is location A or starting position
30 'P2 is the place location
40 MOV PSAFE
```

```
50 HOPEN 1 'the motorized hand is opened
60 MOV P1, -30 'robot is position 30 mm above block
70 MVS P1 'robot moves into position to pick up block
80 HCLOSE 1 'the motorized hand closes
90 MVS, 30 'moves from the current location to a position retracted 30 mm straight up
```

See **Figure 3-2**. Save your program as AC02.prg often during this activity.

Figure 3-2. Example of axis movement commands.

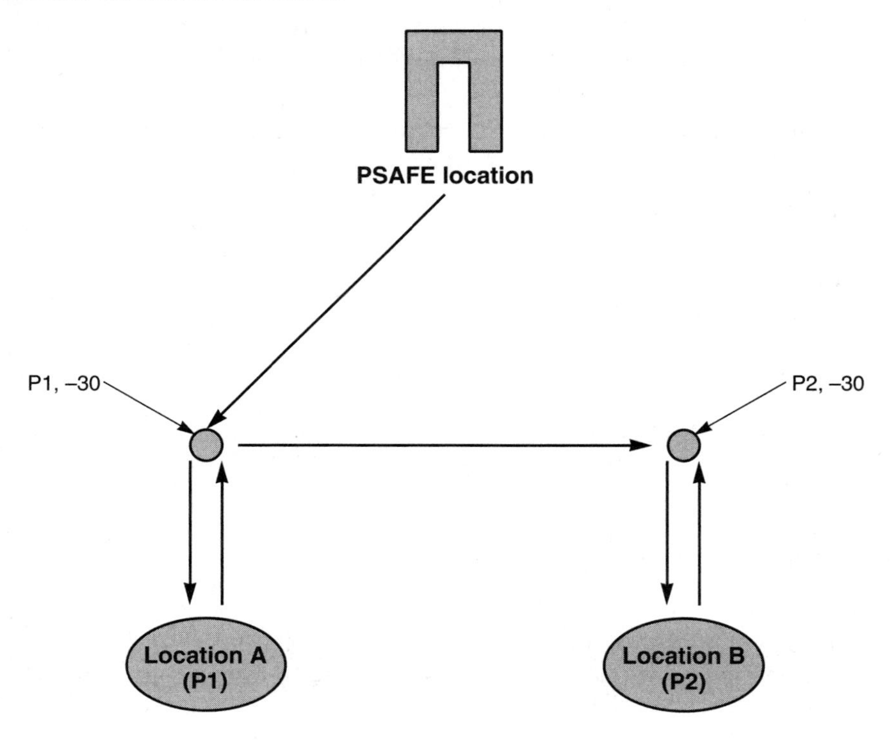

Analysis:

For the robotics software you are using, complete the following.

1. What programming statements (commands) will cause the robot to move to a point location?

 ____

2. Compare the **MOV** and **MVS** commands. When would you use one over the other?

 ____

 ____

3. When is it *not* necessary to use grasp force when using the **HOPEN** and **HCLOSE** commands?

 ____

 ____

4. What is the purpose of the **PSAFE** location?

 ____

5. What is the difference between joint interpolation and linear interpolation?

 ____

Activity 3-3—Circular Interpolation Commands

Name ______________________ Date ______________ Score __________

Objectives:

The purpose of this exercise is to provide information about robotic movements along a three-dimensional arc using three points; a start point, an end point, and a center point.
After completing this exercise, you will be able to:

- Write a program that moves the robot along an arc passing through a transit point.
- Write a program that moves the robot in a circular using three points.

Programming Syntax:

MVR	Move with radius through transit point
MVR2	Moves with radius without passing through transit point
MVR3	Moves from start point to end point with angle less than 180°
MVC	Makes a circle

Examples:

MVR P1, P2 P3	Robot moves with circular interpolation through each point: P1-P2-P3
MVR2 P1, P2 P3	Robot moves with circular interpolation from P1 to P3; P2 is a reference point
MVR3 P1, P2 P3	Robot moves with circular interpolation from P1 to P3; P2 is center point and smallest fan angle used
MVC P1, P2 P3	Robot moves with circular interpolation from P1-P2-P3-P1, creating a circle

Procedure:

Note: The software being used for this lab manual is the MELFA BASIC IV software. The procedure that follows is used to show programming methods.

Using your program from Activity 3-2, create a third point (P3) that becomes your new Location B. Use the **MVR** command with P2 as your transit point for your first drop (the block will be picked up and move along a circular path to get to Location B: P1-P2-P3). At the end of your program (when the block is placed back in its original location of P1), create a new point (P4) that is 50 mm directly above P1 and move the robot in a circle returning to P4 using **MVC** with P5 and P6 (P4-P5-P6-P4). The robot should return to **PSAFE** at the end of program. The following commands should be used in this activity:

- **MVR**
- **MVC**

See **Figure 3-3**. Save your program as AC03.prg often during this activity.

Figure 3-3. Using the **MVC** and **MVR** commands for circular movement.

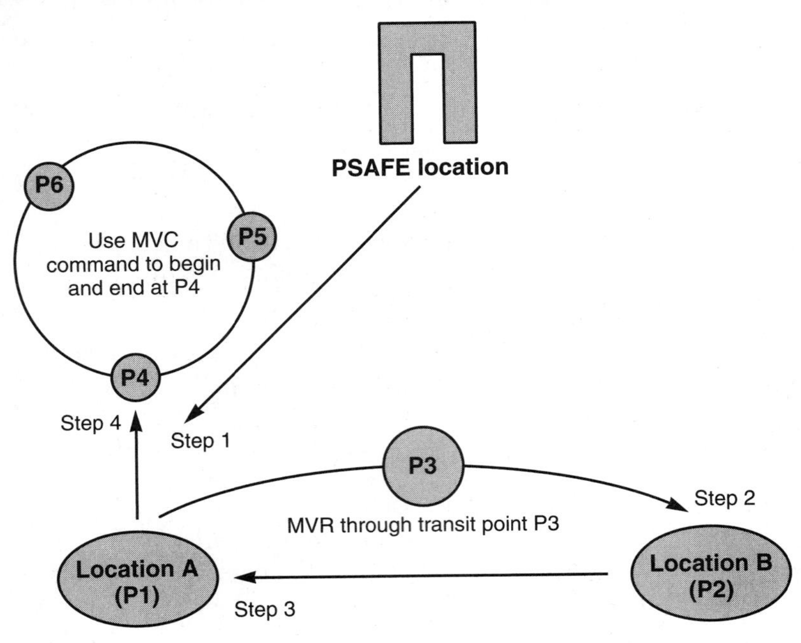

Analysis:

For the robotics software you are using, complete the following.

1. Which programming statement moves the robot in a circular path without passing through a reference point?

   ____

2. Which programming statement returns the robot to its original location to create a circle?

   ____

3. When using an **MVR3** command, what is the second variable in the statement used as?

   ____

   ____

4. When using an **MVR3** command, how does the robot determine the angle or direction to go from the start point to the end point?

   ____

Activity 3-4—Delay or Timer Commands

Name ______________________ Date ____________ Score ________

Objectives:

The purpose of this exercise is to provide information about delaying or halting motions for a specified period of time. One of the most common applications of delay commands is before and after opening and closing the hand to ensure the piece is positioned or cleared properly. Delay commands are often used in industrial applications to pulse an output for a specified period of time to indicate a condition of placement, warning, or errors in a process.
After completing this exercise, you will be able to:

- Write a program that delays opening and closing the robot's end-effector.
- Write a program that incorporates pulsing for use with external devices, such as PLCs or parts counters.

Programming Syntax:

DLY Time delay in millisecond increments

Examples of Programming Syntax:

Time Delay Conversions	
DLY 0.05	50 milliseconds
DLY 0.10	100 milliseconds
DLY 1.00	1 second
DLY 45.00	45 seconds

DLY 1.50 Delay or halt robot movement for 1.5 seconds
M_OUT(10) =1 DLY 1.0 Pulses output 10 for 1 second

Procedure:

Note: The software being used for this lab manual is the MELFA BASIC IV software. The procedure that follows is used to show programming methods.

Using your program from Activity 3-4, add a 500 millisecond (half-second) delay before each **HCLOSE** command and a one-second delay after each **HOPEN** command. After the block is returned, but before the final circular interpolation, pulse output 8 for half a second, or 500 milliseconds.

Save your program as AC04.prg often during this activity.

Analysis:

For the robotics software you are using, complete the following.

1. What is the importance of using a delay command when picking and placing objects?

   ____

2. In what other situations would a delay command be used?

   ____

   ____

3. What is the purpose of pulsing outputs in industrial applications?

   ____

   ____

Activity 3-5—Speed Control Commands

Name ______________________________ Date ______________ Score ____________

Objectives:

The purpose of this exercise is to provide information about controlling the acceleration, deceleration, and speed of the robot's axial and hand movements.
After completing this exercise, you will be able to:

- Write a program that accelerates the robot's movements while traveling to a specified location.
- Write a program that sets the hand end speed while the program is running.
- Write a program that overrides the entire program's speed relative to the robot's maximum speed.

Programming Syntax:

ACCEL	Acceleration/deceleration during movement
OADL	Specifies whether optimum acceleration/deceleration should be set
SPD	Sets the hand end speed in mm/sec
OVRD	Overrides robot's speed applied to the entire program
JOVRD	Sets the joint speed override command

Examples of Programming Syntax:

ACCEL 50, 100	Sets robots acceleration rate to 50% of maximum speed and deceleration rate to 100% Rates are reset to default speeds with reset or by an **END** statement For smooth, constant movements, set rates equal Values greater than 100 will force value 100 Value less than 1 will incur error
OADL ON	Starts optimal acceleration/deceleration The **OADL ON** control takes precedence over an **ACCEL** control
OADL OFF	Ends optimal acceleration/deceleration
SPD 10	Sets hand speed to 10 mm/sec **SPD** valid for robot's linear and circular movements Often used while testing programs to ensure safety of hand end speeds **SPD M_NSPD** used to set hand speed back to default
OVRD 75	Sets the entire program to 75% of the robot's maximum speed **OVRD M_NOVRD** sets the default speed
JOVRD 25	Sets the robot's joint movement to 25% of the robot's maximum speed Valid only during joint interpolation **M_NJOVRD** sets joint speed default value

Procedure:

- Using your program from Activity 3-4, set the entire program's speed to 25%, then to 50%, then to 75%, and back to 100%. Observe the differences in the robot's speed.
- Write a program that will pick up a cylindrical part and place it inside a round hole. Use optimal acceleration to move from the pick location to the place location and a hand end speed of 15 mm/sec when inserting part into the hole. Remember to return the robot to the safe location after the part is released. See **Figure 3-4**. Save your program as AC05.prg often during this activity.

Figure 3-4. Example of how a program uses speed control to pick up a cylindrical object and place it into a hole.

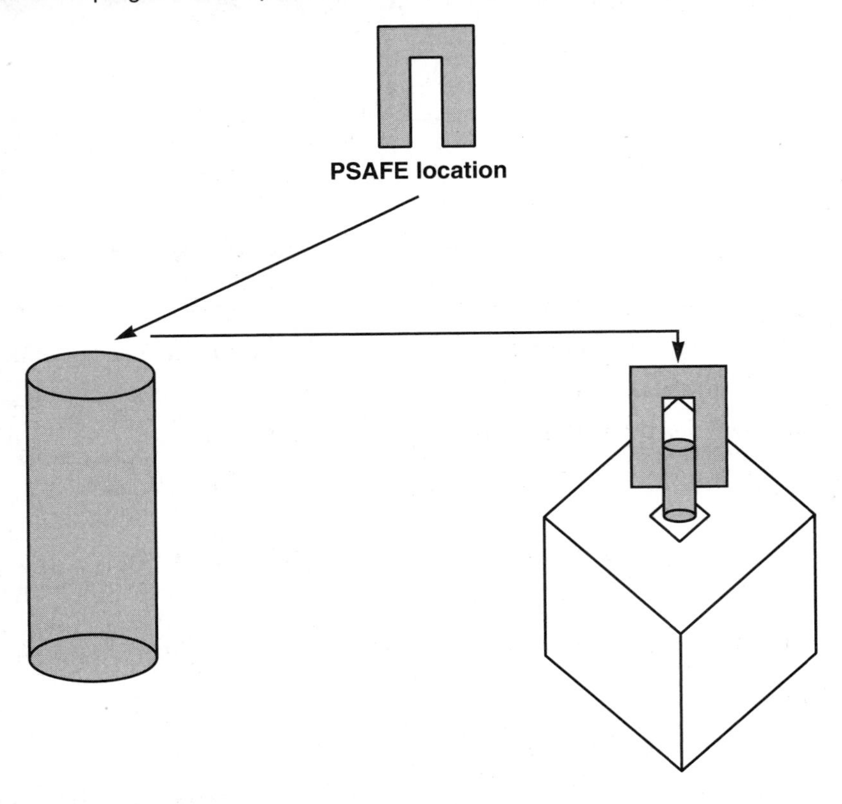

Analysis*:*

For the robotics software you are using, complete the following.

1. Explain why you cannot use optimal acceleration and the **ACCEL** command concurrently.

   ____

2. Which programming command controls the robot's joint movements only?

   ____

   ____

3. In addition to testing programs, what other applications might be good candidates for using **SPD** control?

   ____

   ____

4. What speed will the robot set when an acceleration of 120 is programmed?

   ____

5. What speed will the robot set when an acceleration of .5 is programmed?

   ____

Activity 3-6—Program Control GOTO Function

Name ______________________ Date ______________ Score ____________

Objectives:

The purpose of this exercise is to provide information about the **GOTO** command that allows the program to jump unconditionally to a specified line. It is important to understand that this command does not return automatically to the original line and must be "returned" in some fashion if necessary. The lines that are skipped when the **GOTO** command is invoked are not used during that iteration of the program.

After completing this exercise, you will be able to:

- Use a **GOTO** command successfully to "jump" lines in a program.

Programming Syntax:

GOTO <line number>	Program jumps to a designated line
ON <expression> GOTO <line number>	Program jumps to a designated line according to the conditions of an integer variable

Examples of Programming Syntax:

GOTO 200	Program jumps to line 200
ON M1 GOTO 200	When a variable called M1 = 1, go to line 200

Procedure:

Note: The software being used for this lab manual is the MELFA BASIC IV software. The procedure that follows is used to show programming methods.

Below is the example from Activity 3-2:

```
10 'PSAFE is the safe position out of the way of obstructions
20 'P1 is location A or starting position
30 'P2 is the place location
40 MOV PSAFE
50 HOPEN 1 'the motorized hand is opened
60 MOV P1, -30 'robot is position 30 mm above block
70 MVS P1 'robot moves into position to pick up block
80 HCLOSE 1 'the motorized hand closes
90 MVS, 30 'moves from the current location to a position retracted 30 mm straight up
```

Rewrite the program to invoke a line jump each time the hand opens or closes and jumps back after that process is complete. The benefit of using the **GOTO** command is that the program is simplified by calling up lines of code that repeat at different parts of the program.

Information required to complete this exercise:

- Insert 0.5 second delay commands before and after each open and close command
- Lines 10–60 remain exactly the same
- Line 70 should jump to line 1000 to begin your **HOPEN**/**HCLOSE** and **MVS** (this is where your **GOTO** command should jump to)

Example for picking up part:

```
1000 MVS, -30 'moves down to actual pickup position
1010 DLY .5
1020 HCLOSE 1 'closes gripper
1030 DLY .5
1040 MVS, 30 'moves back up
1050 GOTO 80 'this is where the program will jump back to the original program sequence
```

- Line 1060 should begin placing commands and return to your programming when placing is complete
- Add the **END** command before line 1000
- Complete the entire program
  Save your program as AC06.prg often during this activity.

Analysis:

For the robotics software you are using, complete the following.

1. What must one remember when using a **GOTO** command to jump to a line out of the robot's sequence?

   ____

2. Explain how the **ON GOTO** command adds a level of sophistication and conditions to the program.

   ____

   ____

3. Describe an environment or situation that would require an **ON GOTO** command.

   ____

   ____

Activity 3-7—Program Control If...Then...Else

Name ______________________________ Date ______________ Score ____________

Objectives:

The purpose of this exercise is to provide information about the **IF...THEN...ELSE** command. The command executes specified instructions that correspond to the **THEN** statement when true or **ELSE** statement when false, as they relate to robot or external input/output bits. The function **IF...THEN...ELSE...ENDIF** allows for several lines of code.

After completing this exercise, you will be able to:

- Use an **IF...THEN...ELSE** statement to jump to a specific line based on the **THEN** and **ELSE** statements.
- Use an **IF...THEN...ELSE...ENDIF** statement to make a decision about where to place a part based on predetermined conditions.

Programming Syntax:

IF...THEN...ELSE — **IF** (logical expression) **THEN** (process if expression is true) **ELSE** (process if expression is false)

IF...THEN...ELSE...ENDIF — **IF** (logical expression) **THEN** (process if expression is true) **ELSE** (process if expression is false) **ENDIF** (ends the entire statement)

Examples of Programming Syntax:

```
10 IF IN_(8) = 1 THEN 200 ELSE 400
```

Program jumps to line 200 if M_(8) = 1 or else it will jump to line 400

```
10 IF M1> 10 THEN 20
```

Multiple line statement that evaluates the numeric variable M1. If M1 is greater than 10, execute line 20. If M1 is less than or equal to 10, execute line 30. Lines 20 and 40 set the value of the numeric variable M1 to 10 or –10, respectively.

```
20 M1 = 10
30 ELSE
40 M1 = -10
50 END IF
```

Procedure:

Note: The software being used for this lab manual is the MELFA BASIC IV software. The procedure that follows is used to show programming methods.

Below is the first half of a two-part quality check program that incorporates the **IF...THEN...ELSE** and **IF...THEN...ELSE...END** statements.

```
10 'PSAFE is the safe position out of the way of obstructions
20 'PPICK is the pickup location of each new part
30 'P1 is the 1st quality check location
40 'P2 is the 2nd quality check location
50 'P3 is the "bad parts" bin that defective items go into
60 'P4 is the "good parts" bin ready for the next process
70 MOV PSAFE
80 MOV PPICK
90 MOV P1 'this is the 1st quality check
100 IF M_IN(10) = 1 THEN 110 ELSE 1000 'M_IN(10) refers to an external quality sensor
110 MOV P2 'this is the 2nd quality check
120 IF M_IN(11) = 1 THEN 'the beginning of the IF...THEN...ELSE...ENDIF
...
1000 MOV P3 'robot moves to
1010 GOTO 60 'robot returns to check next part
```

Complete the **IF...THEN...ELSE...ENDIF** portion of the program. If the part is good (**M_(11) = 1**) the robot should drop the part into the "good parts" bin. If the part is not good (**M_IN(11=0)**) the robot should drop the part into the "bad parts" bin. Remember to end the statement with an **ENDIF**. You will also need to include:

- Opening and closing hand procedures for only the following locations: **PPICK**, **P3**, and **P4** (**P1** and **P2** are only quality checks).
- **END** statement at the end of your main cycle.
  See **Figure 3-5**. Save your program as AC07.prg often during this activity.

Figure 3-5. Example of using the **IF...THEN...ELSE** command.

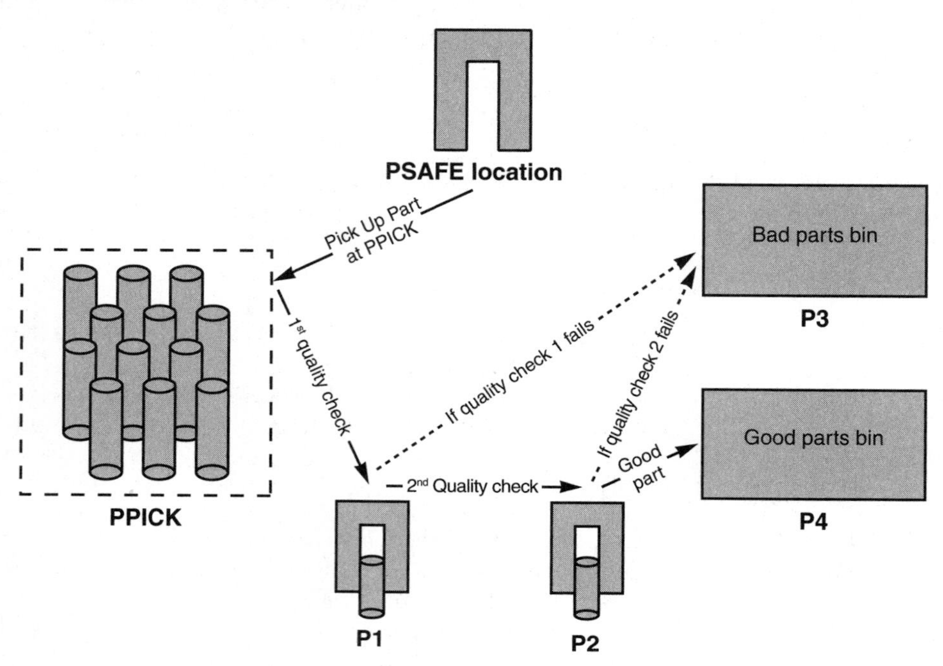

Analysis:

For the robotics software you are using, complete the following.

1. Describe which parts of the **IF...THEN...ELSE** statement correspond to the true and false condition.

   ____

2. Explain how the **IF...THEN...ELSE** command adds a level of sophistication and conditions to the program.

   ____

   ____

3. Describe an environment or situation that would require an **IF...THEN...ELSE** command.

   ____

   ____

4. What is the difference between the **IF....THEN...ELSE** and the **IF...THEN...ELSE...ENDIF** statements?

   ____

   ____

Activity 3-8—Program Control Subroutines

Name ______________________ Date ______________ Score ____________

Objectives:

The purpose of this exercise is to provide information about using the **GOSUB** and **RETURN** commands to call up subroutines. A common application of calling up subroutines is to open and close the grippers. Instead of writing multiple lines of code to (1) move into a pick up position, (2) delay some amount of time before closing, (3) close the grippers, and (4) delay some amount of time before axis movement to make sure part is appropriately grasped, the **GOSUB** statement can call up these program lines as often as needed and be returned to the next line of code for the next commands. Subroutines are placed after the **END** program statement in line numbers that are greater than the **END** statement's line number.

After completing this exercise, you will be able to:

- Use the **GOSUB** and **RETURN** commands to call up opening and closing grippers.

Programming Syntax:

GOSUB	Calls up a subroutine
RETURN	After the subroutine has completed, the **RETURN** statement goes back to the next line of code after the **GOSUB** command

Examples of Programming Syntax:

GOSUB *HOPEN	Goes to the line designated with ***HOPEN**
***HOPEN**	Identifies the named subroutine called **HOPEN**
RETURN	Returns to the next command line after the **GOSUB** statement

Procedure:

Note: The software being used for this lab manual is the MELFA BASIC IV software. The procedure that follows is used to show programming methods.

Using the program from Activity 3-7, change all of your **HOPEN** command lines to subroutines called ***PPICK** and all your **HCLOSE** command lines to ***PPLACE** subroutines. These subroutines will be placed after the program's **END** statement. An example is shown below:

Before

```
10 HOPEN 1
20 MOV P1, -30 'retracted 30 mm above p1
30 MVS P1
40 DLY .5
50 HCLOSE 1
60 DLY .5
70 MVS, -30 'retracts back 30 mm in Z axis
80 END
```

After

```
10 HOPEN 1
20 MOV P1, -30 'retracted 30 mm above p1
30 GOSUB *PPICK 'calls up the PPICK subroutine
40 END
50 *PPICK 'identifies the start of the PPICK subroutine
60 MVS P1
70 DLY .5
80 HCLOSE 1
90 DLY .5
100 MVS, -30
110 RETURN 'returns back to the next line after the GOSUB (Line 40)
```

Save your program as AC08.prg often during this activity.

Analysis:

For the robotics software you are using, complete the following.

1. What is the benefit of using subroutines?

____

2. Explain how a subroutine is initiated in a program.

____

____

3. What will end the subroutine and return it to main program?

____

____

4. Where are the subroutines located in a program?

____

____

Activity 3-9—Program Control Repetition

Name ______________________ Date ______________ Score ____________

Objectives:

The purpose of this exercise is to provide information about using the repetition command (**FOR...NEXT**) as a means of counting. Counting can be done by looping the commands between **FOR** and **NEXT** until they have been repeated set amount of times.

After completing this exercise, you will be able to:

- Use the **FOR...NEXT** command to repeat the statement until the intended conditions are satisfied.

Programming Syntax:

FOR...NEXT	Repeat lines between the **FOR** and **NEXT** statement until conditions are met
FOR...STEP...NEXT	Repeat lines between the **FOR** and **NEXT** at set increments per repetition

Examples of Programming Syntax:

FOR M1=1 to 10
NEXT — Repeat command lines between **FOR** and **NEXT** statements 10 times. The numeric **M1** variable is initially set to 1; increments by 1 for each repetition

For M1=0 to 10 STEP 2
NEXT — Repeat command lines between **FOR** and **NEXT** statements 10 times. The numeric **M1** variable is initially set to 0; increments by 2 for each repetition

Procedure:

Note: The software being used for this lab manual is the MELFA BASIC IV software. The procedure that follows is used to show programming methods.

Observe the following partial program:

```
10 FOR M1 = 1 to 10 'repeats command lines between the FOR and NEXT command 10 times
20 MOV P1
30 MOV P2
40 GOSUB *PPICK 'calls up subroutine called "PPICK"
50 MOV P3 'program picks up here after RETURN statement of PPICK subroutine
60 GOSUB *PPLACE
70 NEXT 'goes back to the FOR command for 10 cycles
80 END
```

Change the program above to repeat 20 times and increment by 2 with each repetition. Save your program as AC09.prg often during this activity.

Analysis:

For the robotics software you are using, complete the following.

1. What is the benefit of using the **FOR...NEXT** command?

   ____

2. Which command syntax would you use when you want to repeat the same lines of a program for a set number of times?

   ____

   ____

3. Which command syntax would you use when you want to incrementally repeat the same lines of a program by a numeric value greater than 1?

____

____

Activity 3-10—Robot I/O

Name ______________________________ Date ______________ Score ____________

Objectives:

The purpose of this exercise is to provide information about interrupts and using signal control to allow your robot to "communicate" with the outside world using its I/O board inside the robot's controller. Although our activity focuses on using switches, sensors, and various other controls directly tied and controlled by the robot, most industrial applications use a Programmable Logic Controller (PLC) to control all external devices. After completing this exercise, you will be able to:

- Use the **M_IN( )** and **M_OUT( )** commands to sense the environment and turn on some output in response to the input.
- Use an interrupt to constantly monitor a condition while executing the robot's program.

Programming Syntax:

DEF ACT	Defines interrupt conditions and process for generating interrupt
ACT	Designates the priority of the interrupt
RETURN	When a subroutine is called up in the interrupt, this command returns to the interrupt source line
WAIT	Waits for an input signal to reach a designated state
M_IN( )	Identifies some input signal
M_OUT( )	Identifies some output signal

Examples of Programming Syntax:

DEF ACT 1, M_IN(10) = GOSUB 50	The subroutine on line 50 is called up when input 10 is turned on
ACT 1 = 1	Enables interrupt #1
ACT 1 = 0	Disables interrupt #1
RETURN 0	Returns to the line where the interrupt occurred
RETURN 1	Returns to the line immediately following where the interrupt occurred
WAIT M_IN(8)=1	Waits for the input signal 8 change to 1 or turn on
M_OUT(10) = 0	Turns the output bit 10 **OFF**
M_OUT(10)=1	Turns the output bit 10 **ON**

Procedure:

Note: The software being used for this lab manual is the MELFA BASIC IV software. The procedure that follows is used to show programming methods.

Observe the following program:

```
10 WAIT M_IN(12)=1 'waits on the condition of input 12 to turn on
20 HCLOSE 2 'closes the pneumatic gripper valve #2 attached to a solenoid
30 DEF ACT 1, M_IN(8) = 1 GOSUB *PPICK 'assigns input 8 to interrupt 1
40 ACT 1=1 'enables interrupt 1
50 MOV P5
60 DLY 3
70 MOV P6
80 MOV P7
90 DLY 0.5
100 MOV P8
110 END
1000 *PPICK
1010 MOV P1
1010 DLY 0.5
1010 MOV P2
```

```
1010 DLY 0.5
1010 MOV PSAFE
1010 HOPEN 2 'opens the valve to reverse solenoid
1010 WAIT M_IN(8)=0 'waits for the part to clear the area before proceeding
1010 ACT 1=0 'disables the interrupt
```

Modify this program by doing the following:

1. Define a second interrupt assigned to signal 1 with a subroutine called ***PCHK**.
2. Enable the interrupt between the current lines.
3. Enable the subroutine before the current line 70.
4. Disable the subroutine after the current line 90.
5. Create a new point in the subroutine **PCHK** where the robot will move and stay there for 5 seconds as a check is done on a part. Then, return back to the next line of code after the interrupt.

Save your program as AC10.prg often during this activity.

Analysis:

For the robotics software you are using, complete the following.

1. What is the benefit of the robot having input and output capabilities?

 ____

2. Why would you choose to use an interrupt command?

 ____

 ____

3. Provide at least three examples of inputs and three examples of outputs that can communicate with a robot.

 ____

 ____

 ____

 ____

4. Why must an interrupt be defined to use the robot's input and output modules?

 ____

 ____

Activity 4-1— Palletizing

Name ______________________ Date ______________ Score ____________

Objectives:

The purpose of this exercise is to provide information about a common method of minimizing the number of positions required to teach the robot, called palletizing. By utilizing the techniques of this activity, the programmer can drastically reduce the time to teach positions, reduce work cell setup time and keep changeover time to a minimum. After completing this exercise, you will be able to write a palletizing procedure to load or unload parts in a rectangular array.

A pallet can be defined as a grid or array of positions that consist of rows and columns with equal spacing between rows and columns. See **Figure 4-1**. Palletizing consists of arranging parts on some surface or space while de-palletizing refers to removing parts from some predetermined surface or space. In spite of the different names, palletizing and de-palletizing are programmed the same, only whether the robot is picking or placing changes.

Figure 4-1. Various examples of pallets.

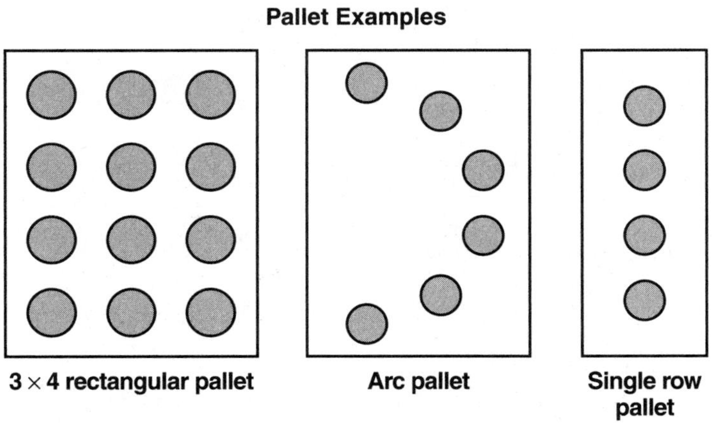

All palletizing procedures using the MELFA BASIC IV software have the following minimum requirements:

- Pallet number
- Start point
- End point A for the pallet row or column (or transit point for arc pallets)
- End point B for the pallet row or column (or transit point for arc pallets)
- Quantity of parts from the pallet's start point to end point A
- Quantity of parts from the pallet's start point to end point B
- Direction of the load/unload sequence

Although three point pallets are more accurate, it is possible to teach four points. Furthermore, there are three directional commands that are used to instruct the robot on how to "navigate" the desired pallet. Pallets are limited to 32,767 parts meaning that the quantity in rows multiplied by the quantity in columns cannot exceed 32,767 parts total. Additionally, the quantity of parts in the rows and columns must be a positive nonzero positive number.

Programming Syntax:

DEF PLT	Defines a 3-point or 4-point pallet
DEF PLT <PLT #>, <Start>, <End A>, <End B>, <Diag.>, <Qty A>, <Qty B>, <Direction>	
PLT #	A constant, 1–8 ,for naming a pallet
Start	The pallet's start point, such as P1
End A	The pallet's end point A, such as P2
End B	The pallet's end point B, such as P3
Diag	The diagonal or fourth point (not used for arc pallets)
Qty A	Number of parts between start and End A
Qty B	Number of parts between start and End B
Direction	Describes how the robot will navigate the pallet (1 = zigzag, 2 = same direction, and 3 = arc)

Examples:

DEF PLT 1, P1, P2, P3, ,4, 3, 1 Defines a 3-point pallet. See **Figure 4-2**. Note the required double commas when not defining a 4-point pallet.

Figure 4-2. Defining a 3-point pallet.

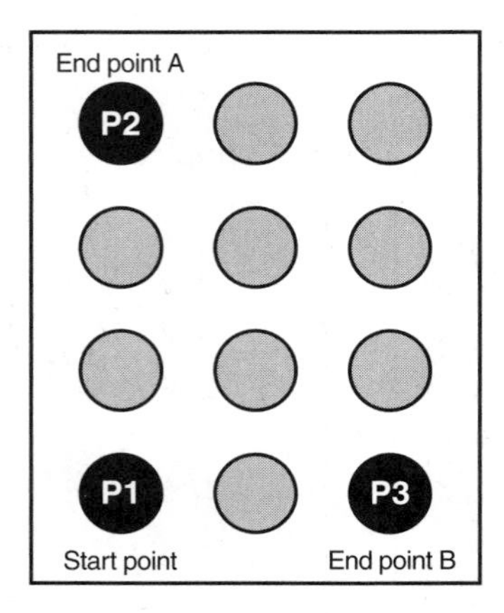

DEF PLT 2, P1, P2, P3, P4, 4, 3, 1 Defines a 4-point pallet. See **Figure 4-3**.

Figure 4-3. Defining a 4-point pallet.

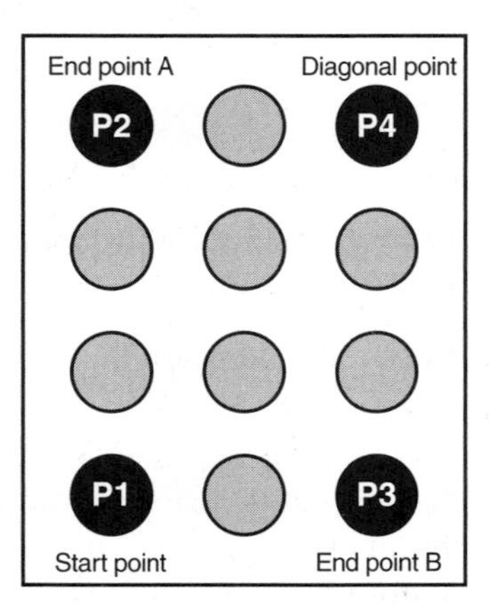

DEF PLT 3, P1, P2, P3, , 5, 1, 3 Defines an arc pallet. See **Figure 4-4**.

Note: This pallet is defined as an arc. The last number is 3. **PLT 3** starts at P1, goes through transit at point P2, and ends at point P3 with five working points.

Name ______________________________

Figure 4-4. Defining a pallet as an arc.

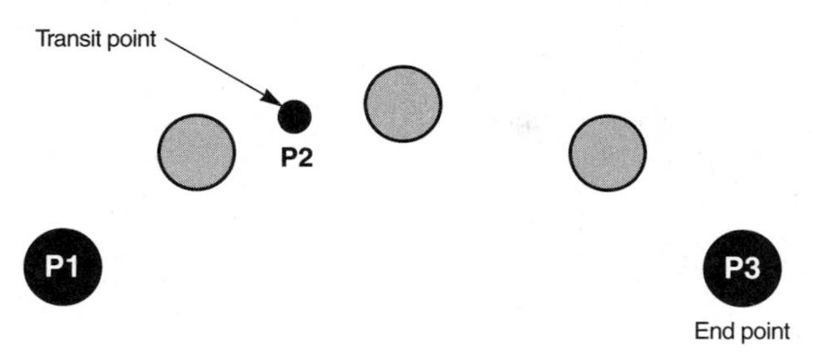

See **Figure 4-5** for a comparison of a zigzag pattern against a load pattern of the same direction.

Figure 4-5. Comparison of load patterns.

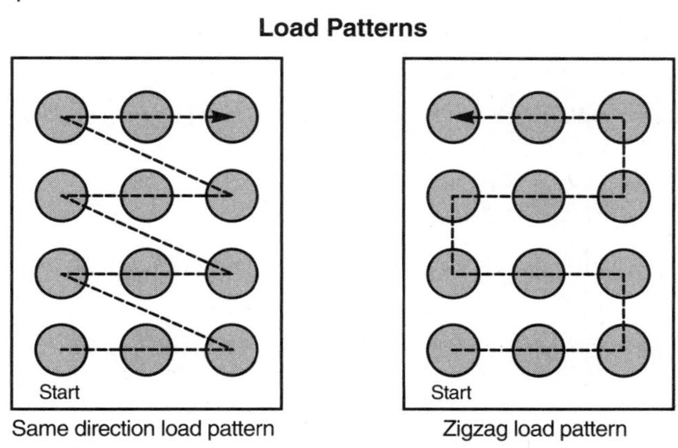

Procedure:

Note: The software being used for this lab manual is the MELFA BASIC IV software. The procedure that follows is used to show programming methods.

Observe the following program:

```
10 'PalletizingExercise
20 'Activity11
30 '2010
40 'Ver01
50 'Pick and Place parts for palletizing
60 'PSAFE position clear of all parts
70 MOV PSAFE
80 GOSUB *OPENG
90 DEF PLT 1,P1S,P1R,P1C,,2,3,1 'first pallet defined 2 x 3 config
100 DEF PLT 2,P2S,P2S,P2C,,1,6,2 '2nd pallet defined 1 x 6 config single row
110 POFSETZ.Z = 60
120 FOR MCOUNT = 1 TO 6 'keeps track of the position
130 PPICK1=(PLT 1,MCOUNT) 'this ties the PLT1 with the position
140 PPICK2=(PLT 2,MCOUNT)
150 MOV PPICK1 + POFSETZ 'moves to the first pick pickup position (see PPICK and P1S below)
160 MOV PPICK1
170 GOSUB *CLOS
180 MOV PPICK1 + POFSETZ
190 MOV PCHK, -35
200 MOV PCHK 'move to check for hole
210 DLY .5
```

```
220 IF M_IN(12) = 1THEN
230 MOV PCHK, -35
240 MRN = MCOUNT 'Sets MRN to the current MCOUNT number
250 WAIT M_IN(15) = 1 'Wait for input 15 on reject
260 GOSUB *DISC
270 ELSE
280 MOV PCHK, -35
290 WAIT M_IN(15) = 1 'Wait for input 15 on reject
300 POFSETZ.Z = 60 'Offset for Z axis
310 MOV PPICK2 + POFSETZ
320 MOV PPICK2
330 GOSUB *OPENG
340 MOV PPICK2 + POFSETZ
350 ENDIF
360 NEXT MCOUNT 'this goes to the next position in the Pallet
370 MOV PSAFE 'Beginning of returning all block to the grid
380 FOR MCOUNT = 6 TO 1 STEP -1 'Palletizes in opposite direction
390 PPICK1=(PLT 1,MCOUNT) 'Sets the current pallet position using MCOUNT must happen on each iteration of for next
loop
400 PPICK2=(PLT 2,MCOUNT)
410 IF MCOUNT = MRN THEN
420 GOSUB *PCKREJ 'If picking up a reject goto reject position to get it.
430 ELSE
440 MOV PPICK2 + POFSETZ
450 MOV PPICK2
460 GOSUB *CLOS
470 MOV PPICK2 + POFSETZ
480 ENDIF
490 MOV PPICK1 + POFSETZ
500 MOV PPICK1
510 GOSUB *OPENG
520 MOV PPICK1 + POFSETZ
530 NEXT MCOUNT
540 MOV PSAFE
550 'HLT
560 END
570 *OPENG
580 DLY 0.2
590 HOPEN 1
600 DLY 0.2
610 RETURN
620 *CLOS
630 DLY 0.2
640 HCLOSE 1
650 DLY 0.4
660 RETURN
670 *DISC
680 MOV PDSC, -10
690 MVS PDSC
700 GOSUB *OPENG
710 MVS, -40
720 RETURN
730 *PCKREJ
740 MOV PDSC, -60
750 MVS PDSC
760 GOSUB *CLOS
770 MVS, -40
780 RETURN
```

Name ____

Modify this program by doing the following:

1. Define Pallet 1 as a 4-point pallet.
2. Pallet one should become a 4 × 3 configuration.
3. Change the quantities in Pallet 2 to reflect the quantities in Pallet 1 in a 2 × 6 configuration.
4. Change all MCOUNTs to reflect the changes in quantities of Pallets 1 and 2.

Save your program as AC11.prg often during this activity.

Analysis:

For the robotics software you are using, complete the following.

1. What is the benefit of using a palletizing command?

   ____

2. Which palletizing procedure would you use when accuracy is a critical factor?

   ____

   ____

3. What is the difference between the two load directions on rectangular pallets?

   ____

   ____

4. How are the distances determined between each point in the rows and columns of a rectangular pattern?

   ____

   ____

Activity 4-2—Control With Dc Stepping Motors

Name ______________________ Date ____________ Score ________

Objectives:

The dc stepping motor represents a type of rotary actuator designed to achieve automatic position control of industrial processing equipment. The motor shaft will move a specific number of degrees with each pulse of electrical energy. The amount of angular displacement produced by each pulse can then be repeated precisely with each succeeding pulse. The resulting output of this motor is used to accurately locate or position a work surface automatically for some manufacturing operation.

In this activity, you will construct a manual switching circuit for a dc stepping motor. This particular motor is capable of moving in 1.8 steps in either direction. The direction of rotation is determined by the switching sequence of the motor coils. Only four coils are utilized in the construction of this motor. As the motor is stepped through a switching sequence, different coil combinations cause the rotor to move to a new location for each step. Only two of the four coils are energized during an operational step.

In Part A of this activity, you will become familiar with the manual switching logic needed to actuate a dc stepping motor to a specific position. You will also reverse the direction of rotation by changing the logic sequence order of the switch combinations. Through this experience, you will be able to observe how switching logic is transformed into rotary motion.

In Part B of this activity, driver transistors are employed as control elements. Toggle switches are used in this case to achieve manual control of the logic sequence through transistor conduction. In an actual stepping motor control circuit, driver transistors are switched by pulses from a logic control center. The next activity is devoted to logic circuitry needed to achieve automatic control of a stepping motor. Remember that dc stepping motors are frequently used in robotic systems.

Equipment and Materials:

- Dc stepping motor (Superior Electric Co. M-111-FD 310 or equivalent).
- 0–15V dc, 1 A power supply.
- 50 Ω, 20 W resistor.
- 470 Ω, 1/8 W resistors (4).
- 1 kΩ, 1/4 W resistor.
- SPST toggle switches (4).
- IN4004 diodes (4).
- 2N3053 transistors (4).
- SPDT toggle switches (2).
- IC circuit construction board.

Procedure:

Part A: Manual Switching Control

1. Construct the manual switching control circuit for a dc stepping motor in **Figure 4-6**.

Figure 4-6. Manual switching control circuit.

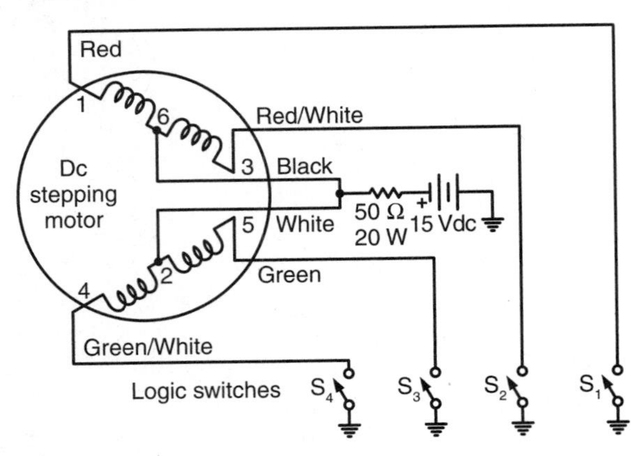

2. Turn on the dc power supply and adjust it to 15V dc before closing any of the logic switches. Then turn off the power supply.
3. Set the logic switches to correspond to Step 1A of **Figure 4-7**. Apply power to the motor by turning on the dc power supply.

Figure 4-7. Logic switch settings.

Step	S_1	S_2	S_3	S_4
1A	on	off	on	off
2	on	off	off	on
3	off	on	off	on
4	off	on	on	off
1B	on	off	on	off

4. Note the direction the rotor moves and its initial location.

   Rotational direction ________________

   You may want to place a strip of masking tape near the rotor shaft and mark its initial location.

   You may need to make a position locating pointer of tape and attach it to the shaft as an indicator of its position change.
5. Set the switches to Steps 2, 3, 4, and finally to Step 1B. Note that Steps 1A and 1B are identical.
6. Starting at Step 1B, set the switches for Steps 4, 3, 2, and 1A in this order.
7. Note the direction of rotation and position of the rotor with reference to its starting location.

   Rotational direction ________________
8. Steps 1A and 1B in practice are referenced as the same step. They were used in this activity as a basis for establishing the rotational direction control logic sequence. They will now be referred to as Step 1.
9. Starting with the switches at Step 1, sequence the Steps 1, 2, 3, 4; 1, 2, 3, 4; 1, 2, 3, 4; and 1, 2, 3, 4.
10. If each step achieves 1.8 of rotation, what is the total rotational movement?

    ____
11. If a protractor is available, you may want to mark the starting position and sequence the switching action and check the final location.

Name ____

12. The rotational sequence is cyclic, and may be started at any position as long as it is maintained in the same logical order.
13. Turn off the power supply. If you have time to do Part B, remove only the switches of the circuit. Otherwise disconnect the entire circuit and return all parts to the storage area.

Part B: Manually Controlled Drivers

1. Connect the manually controlled dc stepping motor with transistor drivers in **Figure 4-8**.

Figure 4-8. Dc stepping motor with transistor drivers.

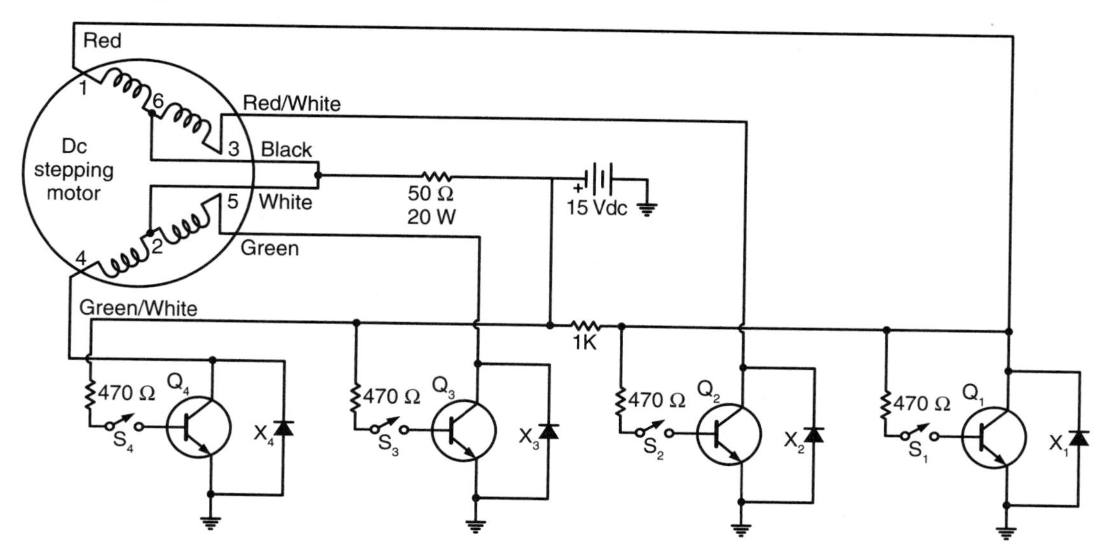

2. The switching action of this motor can be sequenced to turn in either direction as in Part A of the activity. Repeat the stepping sequence of **Figure 4-7** from 1A to 1B. Note the direction of rotation.
3. Reverse the stepping sequence from Steps 1B to 1A and note the direction change.
4. An important aspect of the switching sequence can be seen by carefully studying **Figure 4-7**. Switches 1 and 2, you should note, are always in opposite switching states for each step. The same occurs with switches 3 and 4; when one is "on," the other is in the "off" state. These switches could be easily combined by two SPDT switches for control of all four transistors.
5. Refer to **Figure 4-9** and modify the switches so two SPDT switches achieve control of the four transistors.

Figure 4-9. Transistor circuit controlled by two switches.

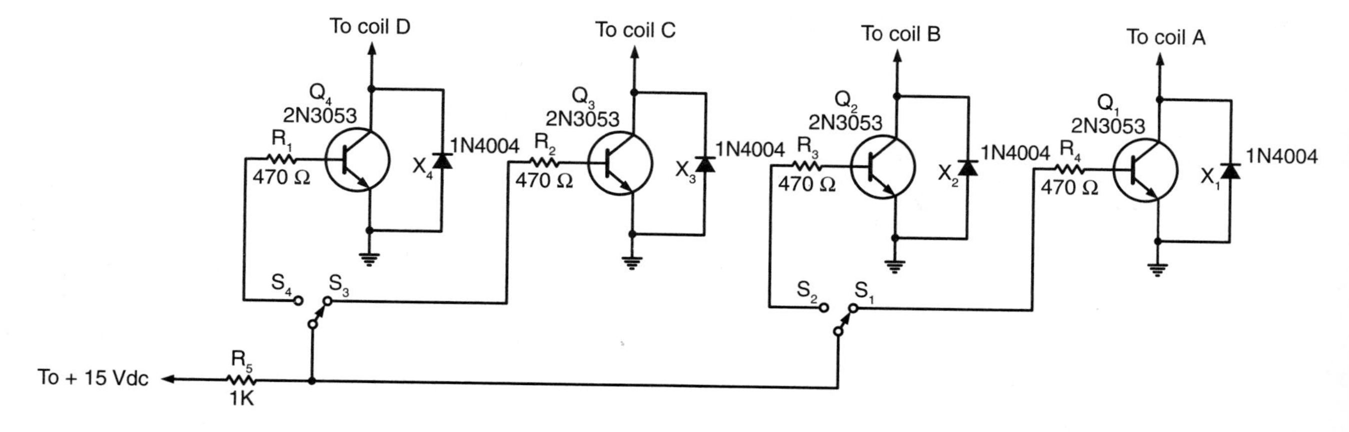

6. When a particular switch setting applies positive voltage to the base of a transistor, it causes conduction. This, in turn, causes the respective motor coil to be energized. The alternate switch position has no bias voltage. Therefore, reverse bias occurs and the respective coil is off.
7. Cycle the motor through Steps 1A to 1B with the two SPDT switches of **Figure 4-9**.
8. Alter the direction of rotation by reversing the switching sequence.
9. It is important to note here, that SPDT switches S_1 and S_2 could be achieved by opposite state outputs of a common flip-flop. This would, of course, permit the switching action of the motor to be achieved automatically.
10. If at all possible, leave the circuit of this activity constructed and go to the next activity. A logic direction sequencing circuit with automatic binary control will be built for the stepping motor.

Analysis:

1. What determines the direction of rotation of a dc stepping motor?

   ____

   ____

2. If a dc stepping motor in Part A were to form a zero resting position 7.2 cw and 14.4 ccw, what would be the logic switching sequence?

   ____

   ____

3. How many sequential steps are needed to produce one complete revolution of the stepping motor used in Part A?

   ____

   ____

4. Why does forward biasing a transistor cause current to the respective coils of a stepping motor?

   ____

   ____

5. How is it possible to control the four transistors with flip-flops?

   ____

   ____

Activity 4-3—Automatic Stepping Motor Control

Name ______________________ Date ______________ Score ____________

Objectives:

Most applications of a dc stepping motor employ some type of control circuit to energize the motor automatically. Manual sequence switching as a general rule is far too slow for typical applications.

Automatic switching circuits are used to control the rotation of a stepping motor. This can be achieved by a common D flip-flop. This type of flip-flop has only one input and two opposite state outputs. See **Figure 4-10** for the D flip-flop circuit.

Figure 4-10. D flip-flop control circuit and truth table.

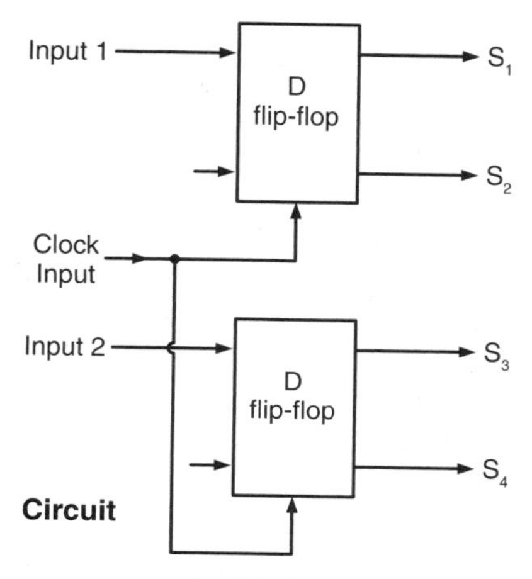

Truth table

Step Number	Input 1	Input 2	Outputs			
			S_1	S_2	S_3	S_4
3	0	0	0	1	0	1
4	0	1	0	1	1	0
2	1	0	1	0	0	1
1	1	1	1	0	1	0

Notice that the outputs of switches 1 and 3 follow the inputs of 1 and 2 respectively. You should also note that output switches 1-2 and 3-4 assume opposite states with respect to their corresponding inputs 1 and 2. This means that four valid switch settings are obtained by using only two inputs to the D flip-flops. Column 1 of the truth table refers to the corresponding step sequence number of **Figure 4-11** of the manual switching circuit.

The combined inputs 1 and 2 form binary numbers ranging from zero to binary three (00-11). If these numbers are used as the inputs for the flip-flops in proper sequence, the motor will rotate by 1.8 steps. **Figure 4-11** shows a summary of step numbers, binary numbers, and the corresponding decimal equivalents for the stepping sequence.

Figure 4-11. Stepping sequence.

Step Number	Binary Number of Switches 1 and 3	Decimal Equivalent
1	11	3
2	10	2
3	00	0
4	01	1
1	11	3

The next function of automatic control is to devise a method by which a proper sequence of binary numbers is developed and applied to the inputs of the two flip-flops. The method employed here is one of a decoding process using NAND gates. See **Figure 4-12**.

Figure 4-12. Circuit for applying binary sequence to flip-flops.

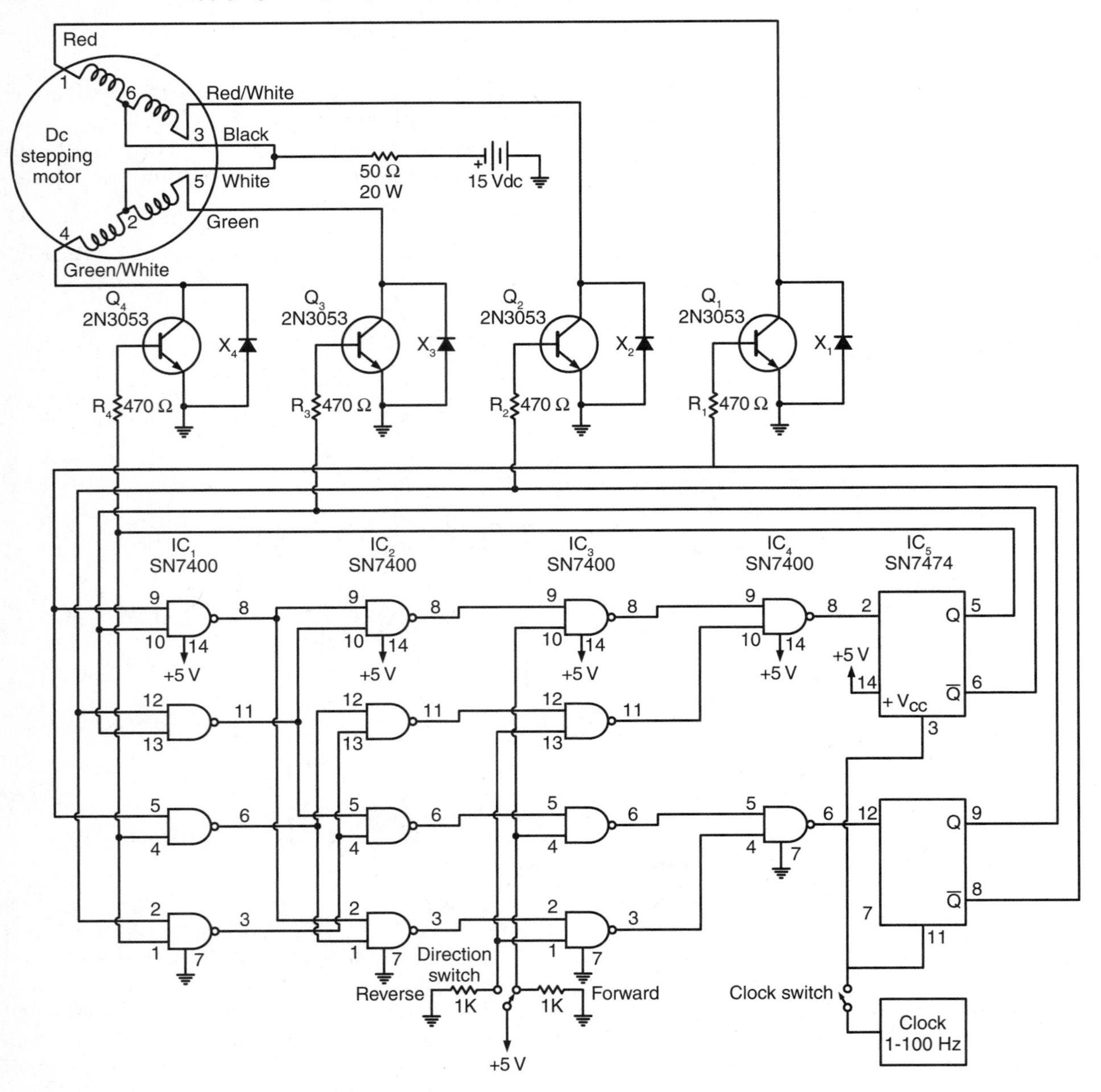

When power is applied to an automatic control circuit, the flip-flops will assume some valid state and apply their output to the driver transistors. This state is sensed by the four NAND gates of IC_1 and the next correct step for either direction is set up by the gates of IC_2. IC_3 then operates in conjunction with the forward/reverse enable switch to determine which of the two possible direction sequences will be applied to the flip-flops. The forward/reverse switch is used to enable or disable each half of IC_3, which influences the direction of rotation.

With power applied to the motor, driver transistors, decoding network, and the direction control switch in either position, stepping does not occur until the flip-flops receive a clock pulse. This means that rotation of the stepping motor is directly dependent on the rate by which clock pulses arrive at the flip-flops. Precision control of the motor is based on the availability of a finite number of clock pulses applied to the flip-flop inputs.

Name ______________________________

In this activity, you will build an automatic directional sequence controller for a dc stepping motor. Through this activity, you will see how a number of ICs are connected together to achieve a specific operational sequence.

You will also be able to test and observe the functional steps of a control circuit in operation. Stepping motor applications of this type are commonly used to control industrial robots.

Equipment and Materials:

- Dc stepping motor (Superior Electric Co. M-111-FD310 or equivalent).
- 0–15V, 1 A power supply.
- 0–5V, 1 A power supply.
- 50 Ω, 20 W resistor.
- 470 Ω, 1/8 W resistors (4).
- 1 kΩ, 1/4 W resistors (2).
- Dc voltmeter or multimeter.
- 2N3053 transistors (4).
- SPDT toggle switch.
- SPST toggle switch.
- SN7400 dual inline ICs (4).
- SN7474 IC.
- IC circuit construction board.
- Clock circuit or generator.
- Oscilloscope.
- Electronic multifunction meter.

Procedure:

1. Construct the dc stepping motor automatic directional control circuit of **Figure 4-12**.
2. Turn on the 5V and 15V power supplies initially with the clock switch off.
3. With a dc voltmeter, measure the driver output signals at the collector of each transistor. When a transistor is conducting, the collector voltage will be low, and when it is nonconducting, it will be approximately 5V. What is the state of the transistor?

   Q_4 ______________, Q_3 ______________, Q_2 ______________, and Q_1 ______________.

   This represents the initial step of the motor.
4. Which step of the truth table of **Figure 4-10** is represented by the transistors?

   Step ______________
5. Set the clock to produce one pulse per second (1 pps) and momentarily turn on the clock switch for one pulse. Repeating Step 3, what is the state of the transistors?

   Q_4 ______________, Q_3 ______________, Q_2 ______________, and Q_1 ______________.

   Which step does this represent?

   Step ______________
6. Repeat Step 5 for two or three steps to determine the sequential pattern of steps. Note that a step change cannot be produced until the clock pulse is applied.

7. Place the direction switch in the reverse enable position. Repeat Steps 5 and 6 to determine the operational sequence. How does it compare with the first readings?

____

____

8. Turn on the clock switch and adjust the clock to produce 1 pulse per second. How does the motor respond?

____

____

9. Increase the clock pulse frequency to 10 pps and observe the change. To stop the motor at a specific position, simply turn off the clock switch.
10. Sequence the stepping operation so the motor moves one half of a revolution and then returns to its original position.
11. Do not disconnect the circuit at this time if you plan to do the next activity.

Analysis:

1. How would the motor respond to 100 or 1000 pulses per second from the clock?

____

____

2. What is the function of the decoder?

____

____

3. Describe the operation of a D flip-flop.

____

____

4. What is the function of IC_3?

____

____

Activity 5-1—Basic Electrical Symbols

Name ______________________ Date ______________ Score ______________

Objectives:

Working with robotic systems requires the ability to read and recognize electrical schematic symbols. Electrical schematics are used to maintain, adjust and repair robotic systems and controllers. Many of these symbols will be used in the activities that follow.

In this activity, you should study and learn to recognize the basic symbols shown in **Figure 5-1**. You may demonstrate your understanding by completing the self-test on electrical symbols.

Figure 5-1. Electrical schematic symbols.

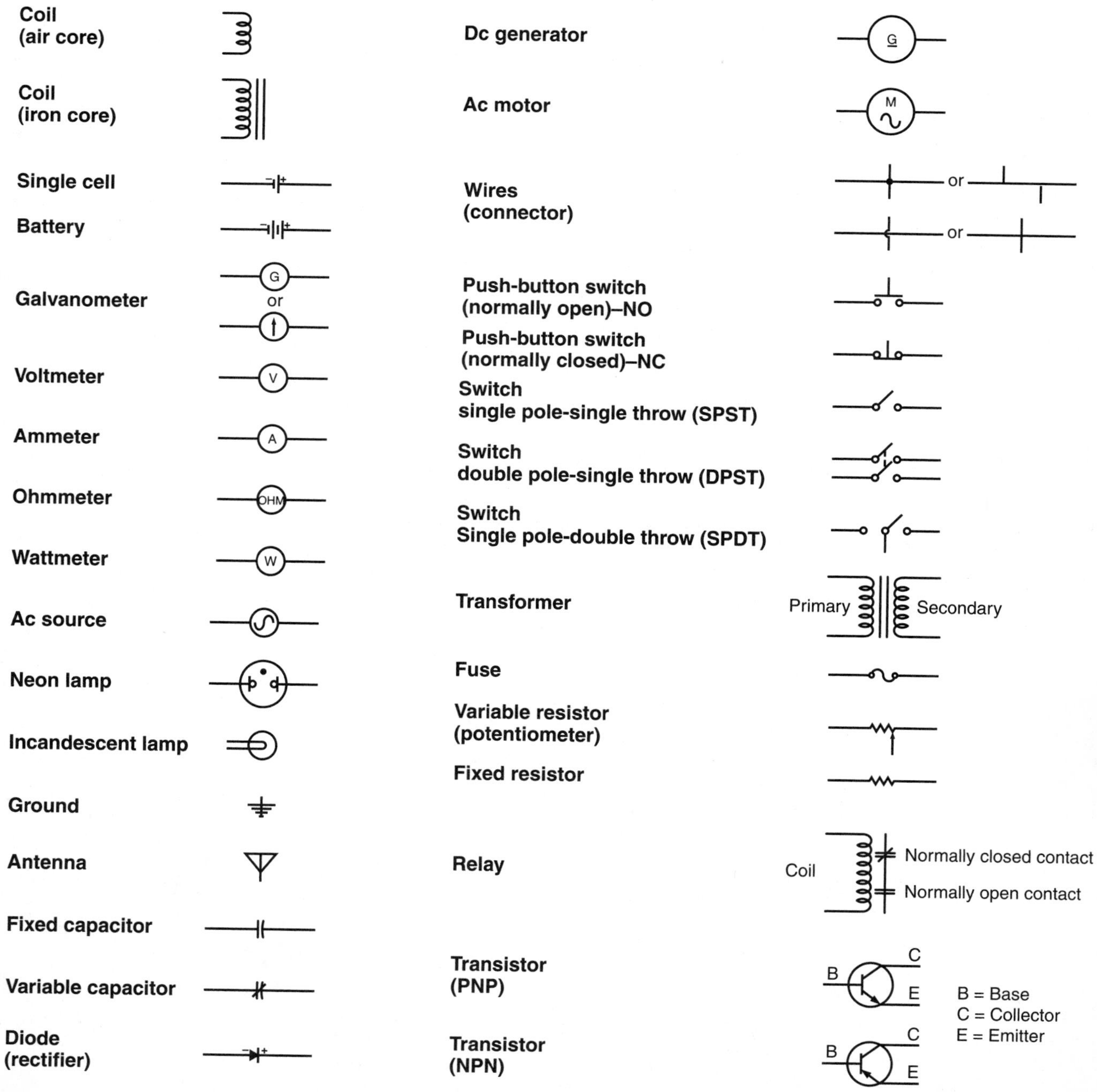

Coil (air core)

Coil (iron core)

Single cell

Battery

Galvanometer (G or ↑)

Voltmeter (V)

Ammeter (A)

Ohmmeter (OHM)

Wattmeter (W)

Ac source

Neon lamp

Incandescent lamp

Ground

Antenna

Fixed capacitor

Variable capacitor

Diode (rectifier)

Dc generator (G)

Ac motor (M)

Wires (connector) (or)

Push-button switch (normally open)–NO

Push-button switch (normally closed)–NC

Switch single pole-single throw (SPST)

Switch double pole-single throw (DPST)

Switch Single pole-double throw (SPDT)

Transformer (Primary, Secondary)

Fuse

Variable resistor (potentiometer)

Fixed resistor

Relay (Coil; Normally closed contact; Normally open contact)

Transistor (PNP) (B, C, E)

Transistor (NPN) (B, C, E)

B = Base
C = Collector
E = Emitter

Analysis:

Without looking at the symbols in **Figure 5-1**, identify each of the symbols that follow. Place the correct response in the space provided.

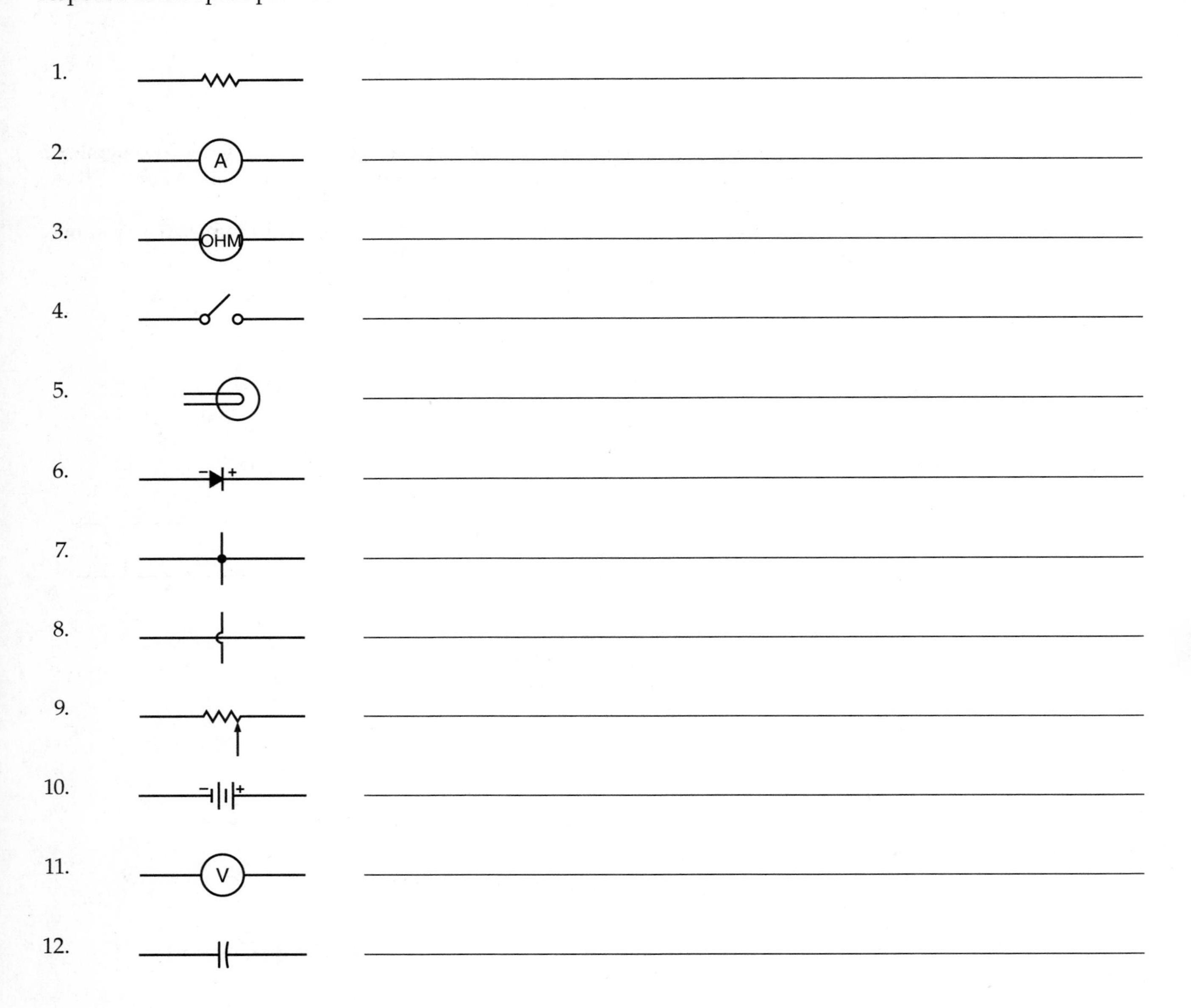

Activity 5-2—Electrical Components, Equipment, and Symbols

Name ______________________ Date ____________ Score ________

Objectives:

Basic to the study of any technical subject is the understanding of the language and symbols used. The study of robotics technology is dependent on the graphic language of electrical and other types of diagrams. Various components, equipment, and symbols are utilized to cause robotic systems to operate properly. You should become familiar with the graphic symbols used in electrical diagrams and learn to recognize the symbols used for common electrical devices.

Equipment and Materials:

- Meter.
- SPST switch.
- Potentiometer (any value).
- Connecting wires for circuit board.
- 6V battery.
- 6V lamp with socket.

Procedure:

1. The wires you will use in this activity are paths for the movement of electrons. Wires may be connected to each other at almost any angle or may cross each other without being connected. The graphic symbols representing these connecting wires, called conductors, are illustrated in **Figure 5-2**.

Figure 5-2. Different schematic symbols for conductors.

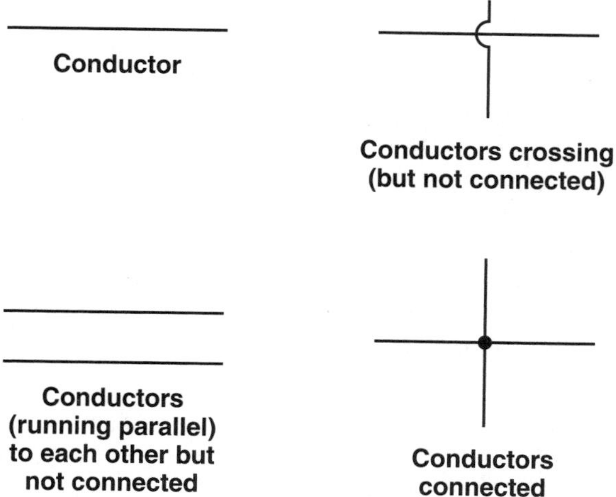

2. In the illustration of **Figure 5-3**, how many times do conductors cross each other and how many times are conductors connected to each other?

   Conductors cross each other ______________ times.

   Conductors are connected to each other ______________ times.

Figure 5-3. Various conductors in a schematic.

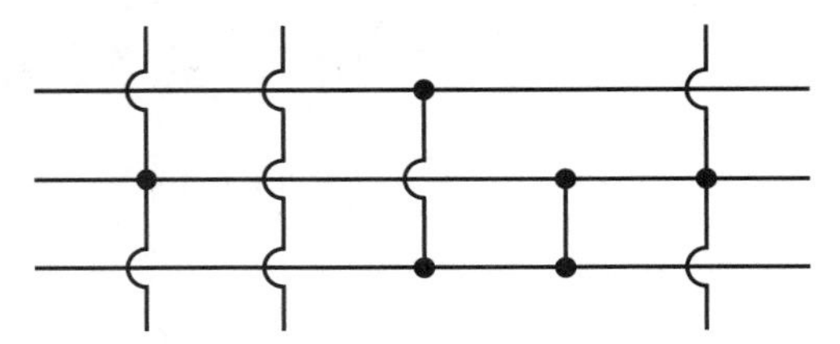

3. A single-pole, single-throw (SPST) switch is a device used to allow current to flow when closed, or on, and interrupt the flow of current when open, or off. A switch is used to turn the light in a room on and off. The symbols representing the SPST switch in its two conditions, on and off, are illustrated in **Figure 5-4**.

Figure 5-4. Schematic symbols for a single-pole, single-throw switch.

4. Fixed resistors are used frequently to control current flow. The symbol for a fixed resistor is illustrated in **Figure 5-6**.

Figure 5-5. Fixed resistor.

5. A light bulb or lamp is a common component used in flashlights and other devices to produce light. The symbol of an incandescent lamp is shown in **Figure 5-6**.

Figure 5-6. Incandescent lamp.

6. A potentiometer, or pot, is a resistive component that can be adjusted to control current flow. A potentiometer is used to increase or decree the volume of a radio or TV set. The illustrations of **Figure 5-7** represent the schematic symbols for a potentiometer. Notice that the center connection of the potentiometer is designated by an arrow and represent the adjustable portion of the device.

Figure 5-7. Schematic symbols for a potentiometer.

7. The 6V battery is a chemical source of electrical energy that causes current to flow through conductors, resistors, and other component. The symbol of **Figure 5-8** is used for a battery. Note that one of the connectors of the battery is labeled with a + sign and the other is labeled with a – sign. It is always important to observe how a battery is connected to a circuit.

Figure 5-8. Battery.

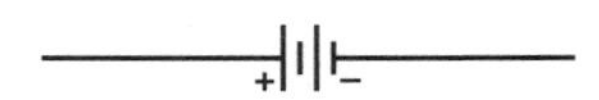

8. Use your circuit board or trainer unit to set up the *circuit* shown in **Figure 5-9**. Several types of circuit boards and trainers that may be used for your experiment are shown in **Figure 5-10**. You should study the layout of the board you are using. Be sure that you connect this circuit and the others you will construct properly. You

Name ______________________________

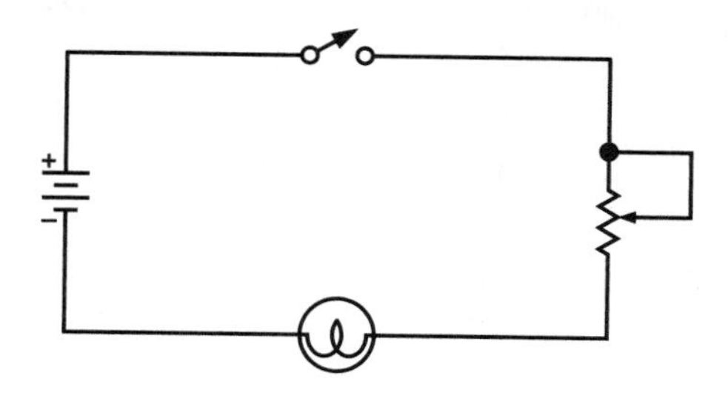

Figure 5-9. Series circuit with potentiometer.

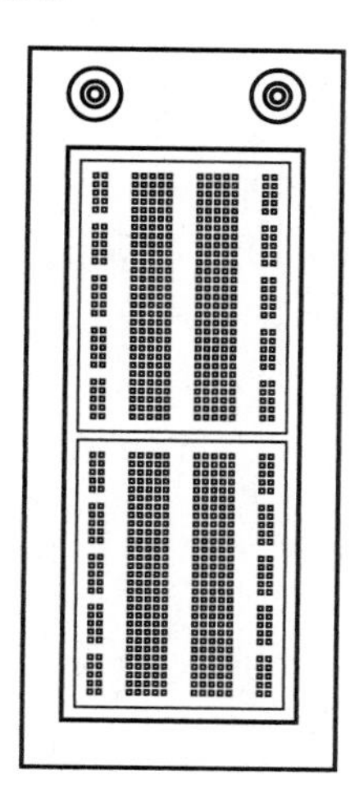

Figure 5-10. Circuit trainer board.

should become very familiar with the use of a circuit board. For most circuit boards, it is necessary to cut wire of the proper diameter to various sizes and strip about 1/4 in. of insulation from both ends. The wires are then used to make circuit connections. The switch and potentiometer should have wires soldered to their terminals about 2 in. long for connecting to the circuit board. They will be used for other experiments.

9. Complete the circuit connection using your board. Close the SPST switch and adjust the potentiometer from its maximum counterclockwise position to its maximum clockwise position. What happens to the lamp? If the circuit is constructed properly, the lamp should get brighter and then dimmer as the potentiometer is adjusted.
10. Adjust the potentiometer until the lamp is at its brightest and open the SPST switch. What happens to the lamp?

____

____

Analysis:

1. Draw the symbols for the indicated components in the space below.

Fixed Resistor   Potentiometer   Open SPST Switch

2. Draw the symbols for conductors in the space below.

   Conductors Crossing　　　　　　Conductors Connected

3. Draw the symbol for a battery, placing the positive (plus) sign and the negative sign (minus) at the proper sides of the symbol.

4. Draw the symbol for a lamp.

Activity 5-3—Electrical Meters

Name ______________________________ Date ______________ Score __________

Objectives:

Electrical meters are used to measure voltage, current, resistance, and other electrical quantities. There are two general types of electrical meters—analog and digital. Analog meters have a hand or pointer that moves according to the amount of the quantity being measured. This type of meter is seldom used today. The other type, digital meters, display numbers indicating the amount of the quantity that is measured. Digital meters are by far the most common type of meter used today. Measurements are made in many types of circuits.

You should learn the proper procedure for measuring resistance, voltage, and current for working with robotic systems. These three measurements are the most common for electrical circuits. Meters used for testing electrical circuits are often referred to as multimeters, such as a DMM (digital multimeter) which is used to measure voltage, current, or resistance. A representative type of digital multimeter will be discussed in this activity.

Measuring Resistance and General Meter Use

Many important electrical tests can be made by measuring resistance. Resistance is opposition to the flow of current in an electrical circuit. The current that flows in a circuit depends on the amount of resistance in that circuit. You should learn to measure resistance in an electrical circuit by using a meter.

A multimeter is used to measure resistance, voltage, or current. The type of measurement can be changed by adjusting the function-select switch (or push button) to the desired measurement. Often, the current range is labeled "mA" (milliamperes) for measuring current, "V" for voltage, and the ohm symbol (Ω) for measuring resistance. Also, a function-select switch or push button is pressed so that the meter may be used to measure alternating current (ac) values or direct current (dc) values.

The ohms measurement function of a DMM is typically divided into ranges, such as 200 Ohms, 2 Kilo-Ohms, 20 Kilo-Ohms and 20 Meg-Ohms. This type of meter is called a multirange meter and there are many different types used. The meter may be adjusted to any of the "Ohms" positions for measuring resistance.

The test leads used with the meter are ordinarily black and red. These colors are used to help identify which lead is the positive or negative side. The positive (red) lead is inserted into the meter for voltage, resistance, and current measurements. The black lead (negative) is used as the "Common" (COM) lead of the meter. Resistance values are sometimes indicated as "K ohms" (multiply by 1000) on the digital scale of the meter for some of the resistance ranges of the meter.

Another type of digital meter is an autoranging meter. The range values do not need to be changed for larger or smaller voltages, currents, or resistances. To measure resistance with this meter, the negative (black) lead is inserted in the "COM" (common) jack and the positive (red) lead is placed in the "V/Ohm" jack. The function select switch of the meter is rotated to the ohm (resistance) symbol. Resistance values are indicated on the digital scale of the meter.

Measuring Voltage

Voltage is applied to electrical equipment to initiate operation. A technician should to be able to measure voltage in order to check the operation of equipment. Many electrical problems develop due to either too much or too little voltage being applied to the equipment. Multimeters are used to measure voltage in an electrical circuit. Typical voltage ranges of a DMM might be 200mV (millivolts), 2V, 20V, 200V and 1000V. When the function-select switch or push button is adjusted to 1000V on the dc volts range, for example, the meter can measure up to 1000 volts dc. The same is true for the other ranges of dc voltage. The voltage value of each range is the *maximum* voltage that may be measured on that range. When making voltage measurements, adjust the function-select switch or push button to the highest range of dc voltage to be measured. The red test lead should be put into the jack labeled "V-Ω-A." The black test lead should be put into the jack labeled "COM." Voltage values are measured with the meter connected *in parallel* with the circuit or component being measured. Voltage values are indicated in volt units on the digital scale of the meter.

To measure voltage with an autoranging meter, the negative (black) lead is inserted in the "COM" (common) jack and the positive (red) lead is placed in the "V/Ohm" jack. The function select switch is placed in one of the voltage settings—"mV" for millivolts, "V" for dc voltage, or "V" (with the ac voltage sine wave symbol above) for ac voltage. Voltage values are usually indicated directly in volts on the digital scale of the meter for each range.

Measuring Current

Current flows through a complete electrical circuit when voltage is applied. Many important tests are made by measuring current flow in an electrical circuit. The current values in an electrical circuit depend on the amount of resistance in the circuit. Learning to use a multimeter to measure current in an electrical circuit is also important.

To measure the current through a circuit, first make sure no voltage is applied to the circuit while you are connecting the meter. Current is measured with the meter *in series* with the circuit. The first step of the procedure for measuring current is to remove a connecting wire from the circuit at the point where the current will be measured. Set the multimeter to the *highest* current range. Connect the negative test lead to the most negative point of the circuit and the positive test lead nearest the positive power source terminal. After the meter has been connected in this manner, voltage may be applied to the circuit. If necessary, adjust the meter to a lower range to get an accurate current reading. The current value should never exceed the value of range setting of the meter. The current value is read directly on the scale of the meter.

Let's look at the controls of a typical multimeter being used for measuring current. The current ranges might be 2 mA (milliamperes), 20 mA, and 200 mA. For example, when the function-select switch or push button is adjusted to 2000 mA on the current range, the meter measures up to 2000 mA (2 amperes). The same is true for the other ranges of current in terms of the maximum current reading. The current value of each range is the *maximum* current that may be measured on that range. When making current measurements, adjust the function-select switch or push button to the highest range of current. The red test lead should be put into the jack labeled "mA." The black test lead should be put into the jack labeled "COM." Current values are usually indicated in milliampere units on the digital scale of the meter.

To measure current with a autoranging multimeter, the range values do not need to be changed for larger or smaller values. The negative (black) lead is inserted in the "COM" (common) jack and the positive (red) lead is placed in the "mA" jack of the meter. The function select switch is adjusted to the "mA" (milliamperes) or "uA" (microamperes) setting, either dc or ac. Current values are indicated in milliamperes or microamperes, corresponding with the setting, on the digital scale of the meter.

The multimeters discussed are representative of the types of digital meters used. The most common type has separate ranges that need to be set for making resistance, voltage, and current measurements. The autoranging meter needs only to be set for the desired function for measuring voltage, current, or resistance.

Analysis:

Obtain a meter to use for completing your lab activities. Study the meter and answer each of the following questions. Indicate "N/A" (not applicable) if the item does not apply for your meter. Explain at the end of the item.

1. What company manufactured the meter?

   ____

2. What is the model number of the meter?

   ____

3. The meter will measure up to __________________ A of dc.
4. True or False: Ohms-adjust control is used each time the resistance range is changed.

   ____

5. To measure dc current greater that 1 A, the range switch is placed in the __________________ position.
6. For measuring current in a circuit, the meter should be connected in (series or parallel) __________________.
7. For measuring voltage, the meter should be connected in (series or parallel) __________________.
8. To measure 18 mA of current, the range switch should be placed in the __________________ position.
9. To measure 10 μA of current, the range switch should be placed in the __________________ range.

10. To measure resistance, the red test lead must be placed in the jack marked __________________ and the black test lead in the jack marked __________________.
11. True or False: Polarity must be observed when measuring ac voltage.

    ____
12. True or False: Polarity is not important when measuring resistance.

    ____
13. Up to __________________ volts dc can be measured with the meter.
14. For measuring a resistor valued at 10 Ω, the __________________ range should be used.
15. True or False: Polarity should be observed when measuring dc voltage.

    ____
16. True or False: It is correct to measure the resistance of a circuit with voltage applied.

    ____
17. True or False: When measuring an unknown value of voltage, one should start at the highest range and work down to the lowest value without exceeding the measured value.

    ____
18. True or False: Meters should always be handled with care and safety.

    ____

Activity 5-4—Industrial Electronic Symbols

Name ______________________________ Date ________________ Score ____________

Objectives:

There are certain electronic symbols that are fundamental to industrial control systems. The majority of the symbols selected for this activity are used frequently in control circuits. You should be familiar with each symbol as a basis of understanding schematic diagrams.

In this activity, you should review the representative symbols in **Figure 5-11** and demonstrate your understanding by completing the analysis.

Figure 5-11. Frequently used industrial electronic symbols.

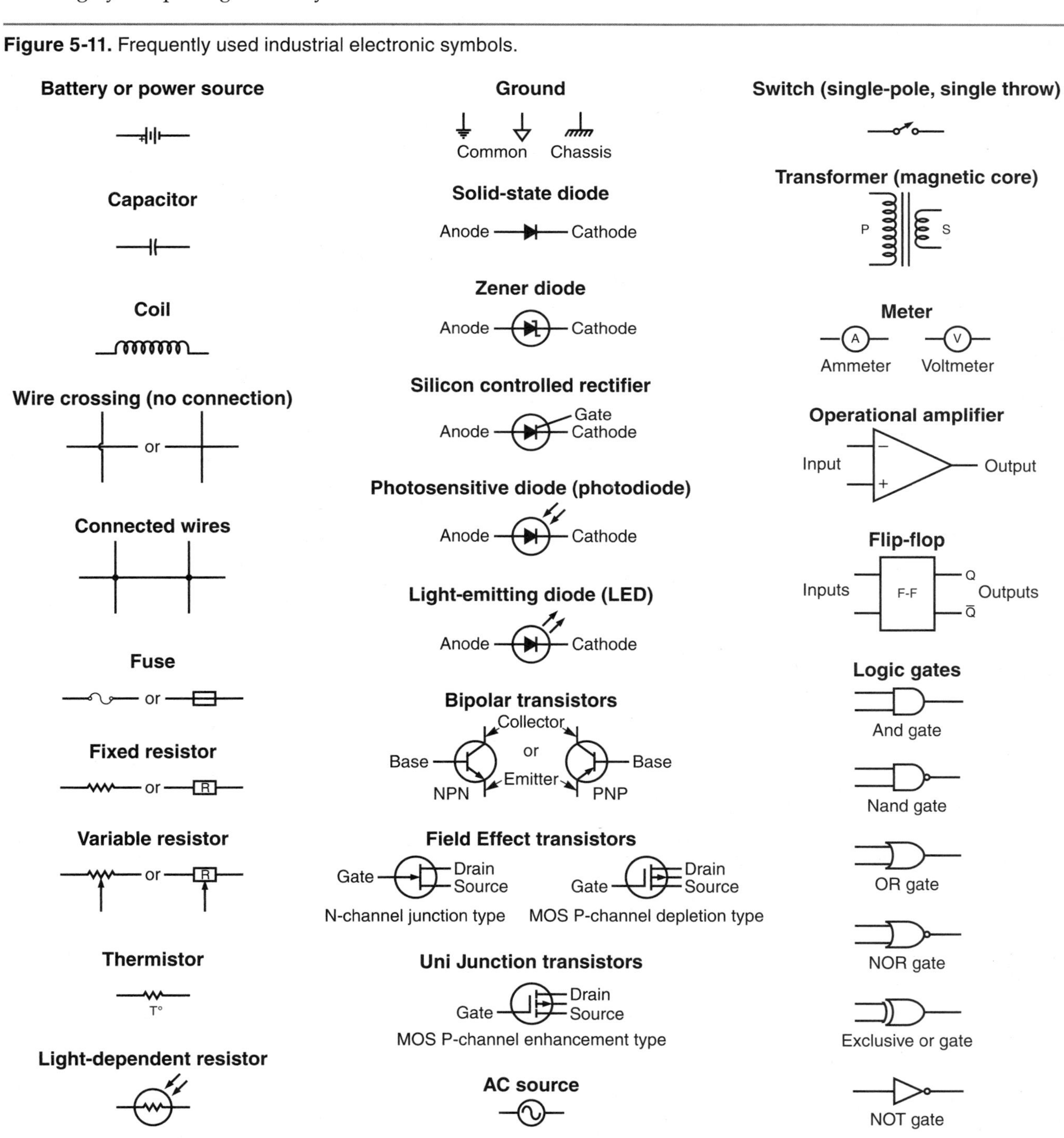

Analysis:

Without looking at the symbols in **Figure 5-11**, identify each of the following symbols. Place the correct response in the space provided.

1. 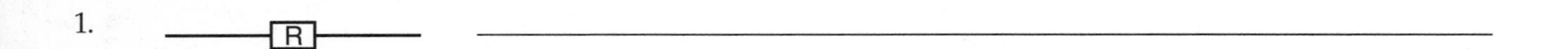

2. 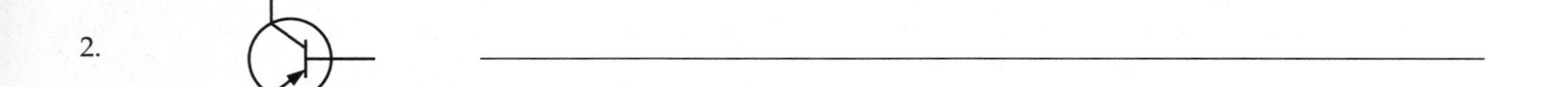

3. 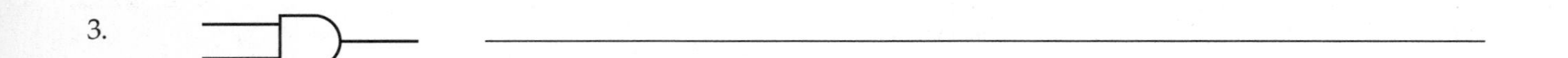

4. _______________

5. 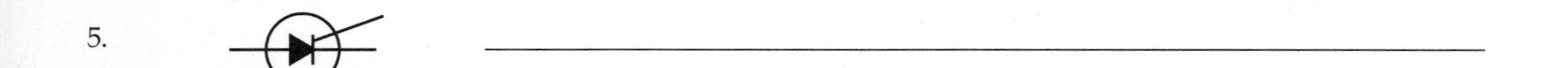

6. 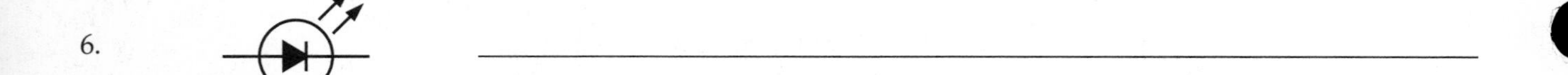

7. 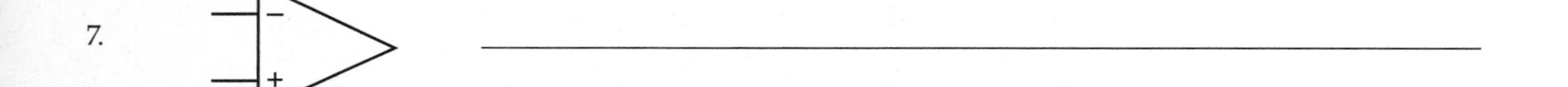

8. 

9. _______________

10. P S _______________

11. _______________

12. + _______________

Name ______________________________

13. 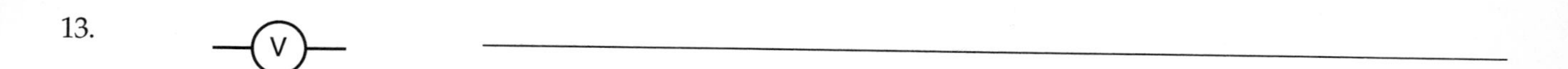

14. 

15. ______________________________

16. 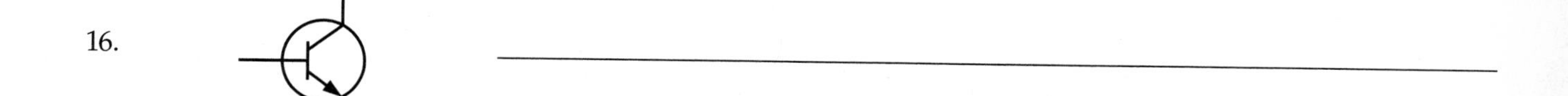

17. ______________________________

Activity 5-5—Basic Electrical Problem Solving

Name ______________________ Date ____________ Score __________

Objectives:

Basic electrical problems are often encountered in any area which involves electrical control systems. Robotics is no exception. The most basic problems in electrical systems involve Ohm's law. Ohm's law is a mathematical formula which explains the relationship between voltage, current, and resistance. This relationship must be understood before electrical concepts are meaningful. In this activity, you will complete some practical problems by applying Ohm's law. Refer to **Figure 5-12** for assistance.

Figure 5-12. Ohm's law circle: *V*, voltage; *I*, current; *R*, resistance. To use the circle, cover the value you want to find and read the other values as they appear in the formula: $V = I \times R$, $I = V/R$, $R = V/I$.

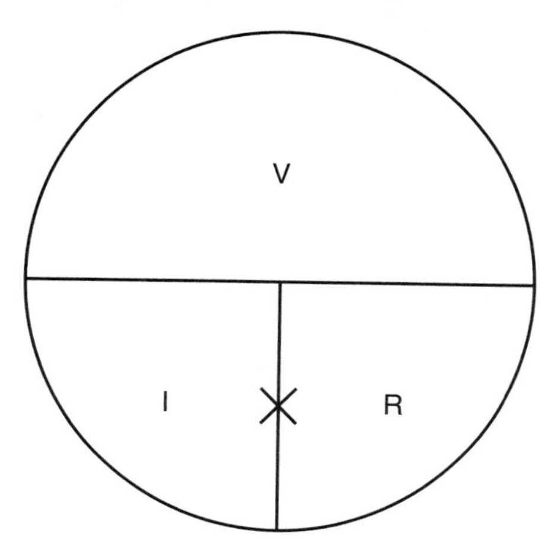

Analysis:

1. A doorbell requires 0.2 ampere of current in order to ring. The voltage supplied to the bell is 120 volts. What is its resistance?

   R = ______________ ohms.

2. A relay used to control a motor is rated at 25 ohms resistance. What voltage is required to operate the relay if it draws a current of 0.25 ampere?

   V = ______________ volts.

3. An automobile battery supplies a current of 7.5 amperes to a headlamp with a resistance of 0.84 ohm. Find the voltage delivered by the battery.

   V = ______________ volts.

4. What voltage is needed to light a lamp if the current required is 2 amperes and the resistance of the lamp is 50 ohms?

   V = ______________ volts.

5. If the resistance of a radio receiver circuit is 240 ohms and it draws a current of 0.6 ampere, what voltage is needed?

   V = ______________ volts.

6. A television circuit draws 0.15 ampere of current. The operating voltage is 120 volts. What is the resistance of the circuit?

   R = ______________ ohms.

7. The resistance of the motor windings of an electric vacuum cleaner is 20 ohms. If the voltage is 120 volts, find the current drawn.

   I = ____________________ amperes.

8. The coil of a relay carries 0.05 ampere when operated from a 50 volt source. Find its resistance.

   R = ____________________ ohms.

9. How much current is drawn from a 12 volt battery when operating an automobile horn of 8 ohms resistance?

   I = ____________________ amperes.

10. Find the resistance of an automobile starting motor if it draws 90 amperes from the 12 volt battery.

    R = ____________________ ohms.

11. What current is drawn by a 5000-ohm electric clock when operated from a 120 volt line?

    I = ____________________ amperes.

12. Find the current drawn by a 50-ohm toaster from a 120 volt line.

    I = ____________________ amperes.

Activity 5-6—Industrial Symbols and Diagrams

Name ______________________________ Date ________________ Score ____________

Objectives:

In Activity 5-1, basic electrical symbols were reviewed. There are several symbols that are unique to industrial applications. In this activity, you will examine some common industrial symbols and diagrams shown in **Figure 5-13**.

Figure 5-13. Industrial schematic symbols.

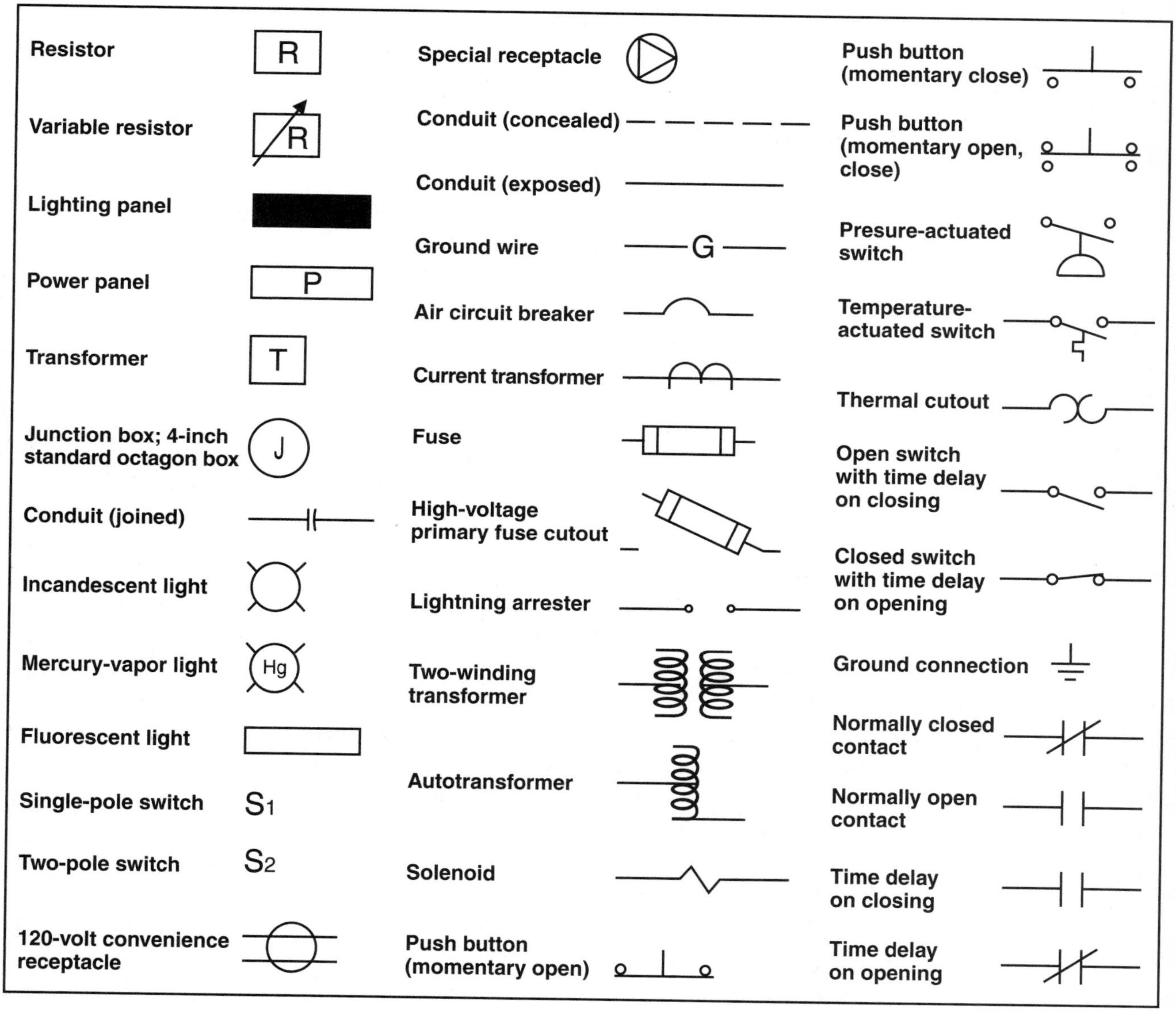

Resistor R

Variable resistor R

Lighting panel

Power panel P

Transformer T

Junction box; 4-inch standard octagon box J

Conduit (joined)

Incandescent light

Mercury-vapor light Hg

Fluorescent light

Single-pole switch S_1

Two-pole switch S_2

120-volt convenience receptacle

Special receptacle

Conduit (concealed)

Conduit (exposed)

Ground wire G

Air circuit breaker

Current transformer

Fuse

High-voltage primary fuse cutout

Lightning arrester

Two-winding transformer

Autotransformer

Solenoid

Push button (momentary open)

Push button (momentary close)

Push button (momentary open, close)

Presure-actuated switch

Temperature-actuated switch

Thermal cutout

Open switch with time delay on closing

Closed switch with time delay on opening

Ground connection

Normally closed contact

Normally open contact

Time delay on closing

Time delay on opening

Industrial Diagrams:

The most common diagrams used to illustrate the components, circuits, and subsystems of an industrial system are connection diagrams, block diagrams, schematic diagrams, and one-line diagrams.

In this activity, you will need to do some research to locate some simple industrial systems that may be represented by using the appropriate type of diagram.

Analysis:

In the space next to each listed device, draw the proper symbol for that device.

1. Incandescent lamp
2. Mercury vapor lamp
3. Fluorescent lamp
4. Circuit breaker
5. Fuse
6. Pushbutton switch (normally open)
7. Pushbutton switch (normally closed)
8. Pressure-actuated switch
9. Temperature-actuated switch
10. Thermal cutout
11. Normally closed contact
12. Normally open contact
13. *Connection Diagram:* The connection diagram shows the physical relationships of the parts of a system. It is used to identify various parts for installation or servicing. The wires that interconnect circuits are ordinarily shown in their correct location. In the following space, draw a simple connection diagram for an industrial system:

Name ____________________

14. *Block Diagram:* The block diagram is a very general method used to show low subsystems of any industrial system fit together to form a functional unit. Rectangles are used to represent subsystems and arrows are used to connect the blocks and show the interrelationships of subsystems. In the following space, draw a block diagram of an industrial system:

15. *Schematic Diagram:* The schematic diagram is used to show the electrical circuit relationships of the various components of an operational system. The electrical, but not necessarily the physical layout of the circuit is shown. This type of industrial diagram is the most often used. In the laboratory activities that follow, this type of diagram will be used. In the following space, draw a schematic diagram of an industrial circuit:

16. *One-Line Diagram:* The one-line diagram is a simplified form of the schematic diagram. This type of diagram is used to show how the components of a system fit together. It is similar to the block diagram in several respects; however, actual symbols, rather than rectangular shapes are used to show component parts. In the following space, draw a one-line diagram of an industrial system:

Activity 5-7—Control Systems Overview

Name ______________________ Date ____________ Score ____________

Objectives:

The term *system* is commonly defined as an organization of parts that are connected together to form a complete operating unit. In this respect, it can apply to a variety of different systems used by industry to achieve some operation in the manufacturing process. An application of this term is the industrial electronic system. Further application of this term includes a wide variety of subsystems under the heading of industrial electronics. Opto-electronic systems, timing systems, digital systems, and environment control systems are all included in this study of industrial systems.

Each industrial system includes a number of unique features or characteristics that distinguishes it from other systems. More importantly, however, there are a number of basic functions common to all electronic systems. Energy source, transmission path, control, load, and indicator are terms commonly used to describe these basic system functions.

In this activity, you will have an opportunity to study a representative system and pick out the basic parts of the system. By doing this, you will become more familiar with the systems concept and be able to apply it to a specific situation.

Procedure:

1. Briefly define the function of the following basic system parts. Use general terms that would apply to all basic industrial electronic systems.

   Energy source.

   Transmission path.

   Control.

   Load.

   Indicator.

2. Explain how the systems concept could be applied to an electrical power system used to supply an industrial plant site.

3. How could the systems concept be applied to the simple electrical circuit of **Figure 5-14**?

Figure 5-14. Electrical circuit.

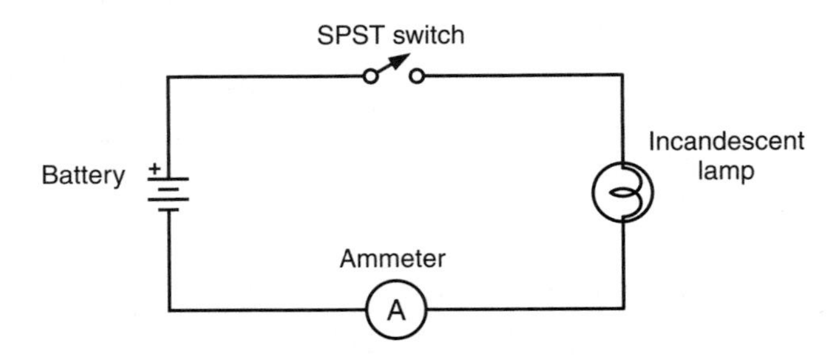

4. The photoelectric box counter in **Figure 5-15** is a typical industrial system. What are the basic parts of this system?

____

____

Figure 5-15. Industrial electronic system.

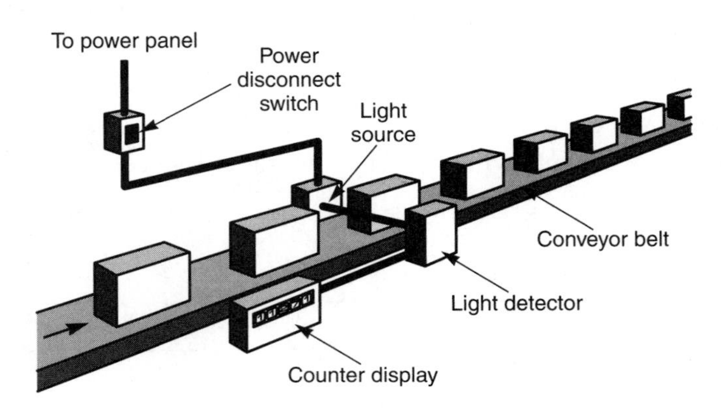

Analysis:

1. Of what significance is the basic systems concept in industry?

____

____

2. What is an opto-electronic system?

____

____

3. What are some applications of an industrial timing system?

____

____

4. What is meant by the term *digital system*?

____

____

5. What are some typical applications of an industrial environment system?

____

____

Activity 5-8—AC Synchronous Motors

Name ______________________________ Date ______________ Score ____________

Objectives:

The ac synchronous motor used in robotic systems is commonly classified as a constant-speed device. It has extremely rapid starting, stopping, and reversing characteristics. Motors of this type only necessitate a simple clockwise-stop-counter-clockwise rotary switch for control. The starting and running currents of a synchronous motor are identical, which is unique for ac motors. This characteristic means that a motor of this type can withstand a high inrush of current when direction changes occur. As a general rule, the synchronous motor can even be stalled without damaging the motor.

Motors of this type are commonly used as drive mechanisms for robotic systems in machinery operations. Starting begins within 1 1/2 cycles of the line frequency, and stopping occurs in five mechanical degrees of rotation. This motor represents a unique part of all servomechanisms used in precision rotary control applications today. In this activity, a simple test circuit will be constructed so you may observe an ac synchronous motor's starting, stopping, and current flow characteristics.

Equipment and Materials:

- Ac synchronous motor—Superior Electric Type SS150 or equivalent motor rated at no more than 200 in./oz.
- Capacitor—3.75 μF, 330V ac.
- Resistor—250 Ω, 25 W.
- SPST switches (3).
- Isolated ac power source—120V.
- Ac ammeter—0–5 A.
- Piece of wood.

Safety:

Use protective eyewear during this activity.

Procedure:

1. Construct the ac synchronous motor test circuit of **Figure 5-16**. Ensure that the motor is mounted securely.

Figure 5-16. Ac synchronous motor test circuit.

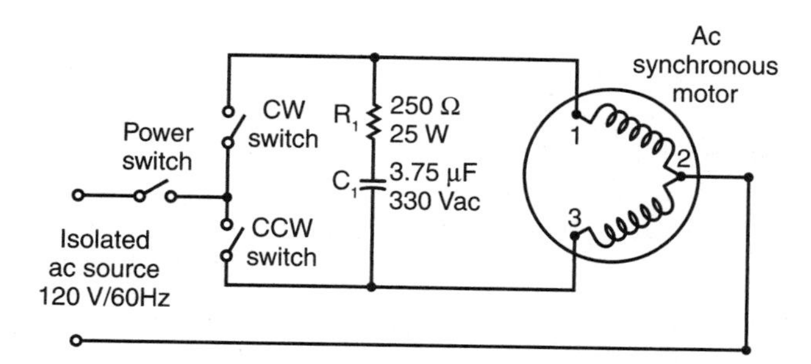

2. Turn on the "cw" switch for clockwise rotation.
3. Momentarily turn on the power switch and observe the rotation of the motor.
4. With a tachometer, measure the speed of rotation.

   Speed = ____________________ RPM.

5. Momentarily turn off the power switch and observe the stopping action of the synchronous motor. You may want to run several trials to see the quickness of the stopping action. How does this compare with other ac motors?

____

6. Note each time that the motor is turned on, how quickly it comes to speed.
7. With an ac ammeter, measure the running current and starting current of the synchronous motor.

   Running current = ________________ amperes.

   Starting current = ________________ amperes.
8. Carefully wedge a piece of wood between the rotating shaft and the bench while holding the motor. This will provide a simple loading method for test purposes.
9. When the motor is loaded down, how does the running current respond?

   ____
10. If the motor is completely stalled, how does the running current respond?

    ____
11. Remove the load from the motor, turn off the clockwise switch, and turn on the counter-clockwise switch. The motor should rotate equally well in the counter-clockwise direction and have the same basic characteristics. Test these again to verify the theory.
12. Switch off the counter-clockwise switch, then switch on the clockwise switch and notice the ease with which direction change occurs.
13. Turn off the ac power source and disconnect the circuit. Return all parts to the storage area.

Analysis:

1. Discuss the operation of an ac synchronous motor.

____
____
____
____
____
____
____
____
____
____
____
____
____
____
____
____

Activity 5-9—DC Stepping Motors

Name ______________________________ Date ______________ Score ____________

Objectives:

Dc stepping motors represent a unique electromechanical rotary actuator that is used in robotic systems. These motors are designed to change electrical pulses into rotary motion. The amount of rotary movement or angular displacement produced by each pulse is repeated precisely for each succeeding pulse. The resulting rotation of these motors may be used to accurately locate or position worktables or fixtures for automatic machining operations. Very precise degrees of accuracy can be achieved with these devices.

In this activity, you will construct a simple test circuit for a dc stepping motor. You will be able to produce very accurate shaft movements by different switching step combinations. Both the four-step and eight-step switching combinations will be tested. Through this activity, you will be able to observe the physical operation and test the accuracy of motor shaft rotation. The switching operation of this activity is achieved automatically by logic devices in actual operating circuits. Precise switching operations obviously improve the accuracy of this circuit.

Equipment and Materials:

- Dc stepping motor—Superior Electric Type M061-FD02 or equivalent.
- SPST switches (4).
- Resistor—50 Ω, 20 W.
- Resistor—1000 Ω, 1/2 W.
- Dc power supply—0–25V, 1 A.
- Diodes—1N4004 (4).

Safety:

Use protective eyewear during this activity.

Procedure:

1. Construct the stepping motor test circuit of **Figure 5-17**.
2. The physical movement of the stepping motor shaft is small and rather difficult to observe. By attaching a simple indicating device such as a small wire or paper pointer to the motor shaft, the mechanical rotation can be readily observed.
3. The switching sequence in the full-step mode will produce 1.8 steps for each switching combination.
4. Adjust the dc power supply to 15 volts. Set the indicator to a starting position and mark it as the starting reference. Then turn on the switch combinations for Step 1 of the full-step mode from the chart in **Figure 5-17**.
5. Go to the next switching combination indicated by Step 2 of the chart. Each combination will produce a rotary step.
6. Follow the stepping procedure through at least two of the four-step cycles. Notice that the switching step combination repeats itself after Step 4.
7. Mark the indicator location as the stopping position for reference.
8. Calculate the rotational degrees by multiplying the number of steps by 1.8.
9. If a protractor is available, measure the number of degrees of rotation between the starting indicator mark and the last switching location mark of the motor shaft.

Figure 5-17. Dc stepping motor test circuit and stepping tables.

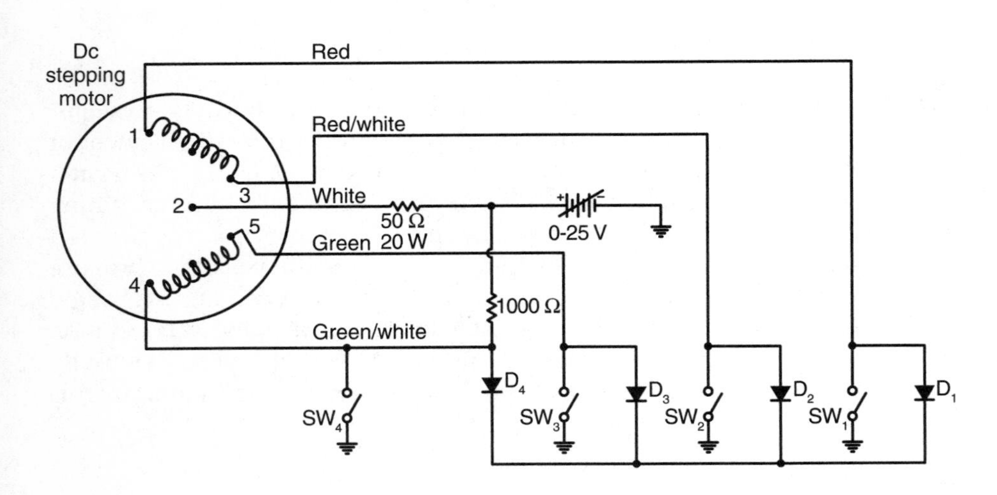

Four-step input (full-step mode)

Step	SW_1	SW_2	SW_3	SW_4
1	On	Off	On	Off
2	On	Off	Off	On
3	Off	On	Off	On
4	Off	On	On	Off
1	On	Off	On	Off

Eight-step input (half-step mode)

Step	SW_1	SW_2	SW_3	SW_4
1	On	Off	On	Off
2	On	Off	Off	Off
3	On	Off	Off	On
4	Off	Off	Off	On
5	Off	On	Off	On
6	Off	On	Off	Off
7	Off	On	On	Off
8	Off	Off	On	Off
1	On	Off	On	Off

10. Repeat the procedure outlined in Steps 2 through 9 of this activity for the eight-step sequence. In this stepping mode, each step produces 0.9 of rotation.
11. Disconnect the circuit and return all components to the storage area.

Analysis:

1. Explain how a dc stepping motor operates?

____

____

____

2. How many steps are required by a dc stepping motor in the full-step mode to produce one revolution?

____

3. How many steps are required by a dc stepping motor in the half-step mode to produce one revolution?

____

4. Describe a typical application of the dc stepping motor.

____

____

____

Activity 5-10—Solenoids

Name ______________________________ Date ______________ Score __________

Objectives:

A solenoid is an electromagnetic coil with a movable core constructed of a magnetic material. The core or plunger is sometimes attached to an external spring. This spring causes the plunger to remain in a fixed position until moved by the electromagnetic field created by current applied to the coil. This external spring also causes the core or plunger to return to its original position when the coil is de-energized.

Solenoids are used for a variety of applications. For example, most door chime circuits use one or more solenoids to cause the chime to sound. Many gas and fuel oil furnaces use solenoid valves to automatically turn on or off the fuel supply on demand. Most dishwashers use one or more solenoids to control the flow of water. The solenoid is representative of the many types of electromechanical control devices used in conjunction with robotic systems.

In this activity, you will learn the basic operation of the solenoid by observing a solenoid as it is used to operate door chimes.

Equipment and Materials:

- Digital multimeter.
- Door chimes (low voltage, 6–24V).
- Variable dc power supply.
- Air-core coil (200 turns, No. 24 wire) or equivalent.
- Cold-rolled steel core, 3/4″ diameter, 6″ long or suitable substitute.
- SPST switch.
- Connecting wires.

Procedure:

1. Adjust the multimeter to measure resistance. Measure and record the resistance of the air-core coil to be used in this activity.

   ______________ Ω

2. Connect the circuit shown in **Figure 5-18**.

Figure 5-18. Solenoid test circuit.

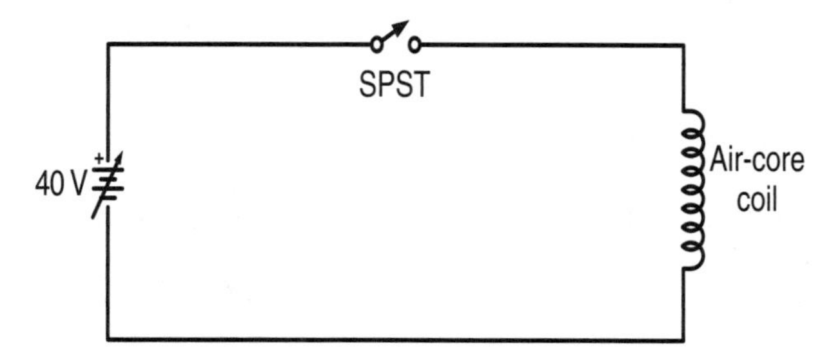

3. Place about 2″ of the end of the cold-rolled steel bar into the coil. See **Figure 5-19**.

Figure 5-19. Air-core coil and steel core.

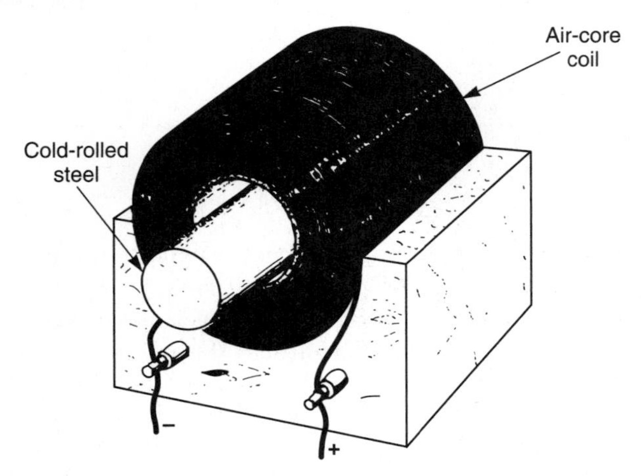

4. Close the SPST switch and describe what happens to the cold-rolled steel core. To prevent the coil from heating, do not allow the SPST switch to remain closed for over 20 seconds.

 ____

 ____

5. Place the bar inside the coil and close the SPST switch. Try to remove the bar with the power on. Describe what happens. Do not allow the coil to remain energized for more than 20 seconds.

 ____

 ____

6. How could the above solenoid be used as a pusher or puller coil?

 ____

 ____

7. Disconnect the circuit in **Figure 5-18**.
8. Acquire a door chime assembly. Examine it to determine if it is to be powered by 6, 12, 18, or 24 volts. Indicate the voltage used for the chime assembly in the space below.

 ______________ V

9. Remove the protective chime cover to expose the spring-loaded solenoids and solenoid connections. Measure and record the resistance of the solenoid coil in the space below.

 R = ______________ Ω

10. There should be three solenoid connections. One will be common to the other two. Connect the common connection and either of the remaining connections to the power supply as shown in **Figure 5-20**.

Figure 5-20. Solenoid circuit from **Figure 5-18** with door chime.

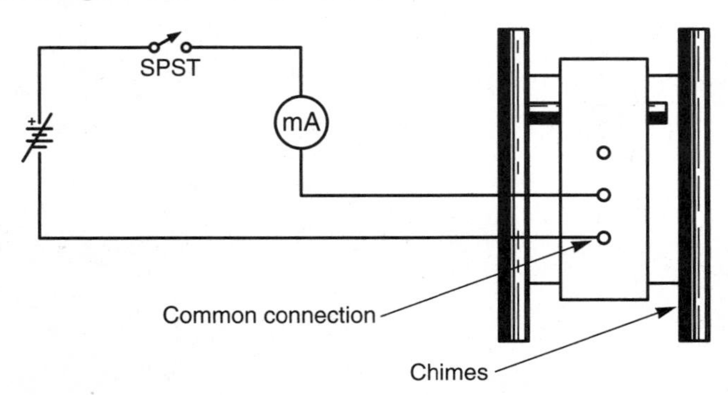

Name ______________________________

11. Close the SPST switch and slowly adjust the power supply from zero until the plunger of the solenoid is drawn into the core. Record the current necessary to pull the plunger.

    ____________ A

12. Open the SPST switch, adjust the power supply to the voltage indicated in Step 8.
13. Open and close the SPST switch several times. Describe what happens.

    ____

    ____

14. Disconnect the circuit and return all components to the storage area.

Analysis:

1. What is a solenoid?

   ____

   ____

2. Where are solenoids used?

   ____

   ____

3. Why are solenoids sometimes called pusher coils?

   ____

   ____

4. Why are solenoid plungers spring-loaded?

   ____

   ____

Activity 5-11—Motor-Driven Timers

Name ______________________________ Date ______________ Score ___________

Objectives:

The motor-driven timer provides a wide variety of timing actions for industrial circuit applications. In its simplest form, this timer has an electric drive motor, ratchet release coil, and a ratchet dial that is held stationary until released. When the timing cycle reaches its set-time, the ratchet is released and the dial resets itself by spring action. Both on delay and off delay reset timers are available. More sophisticated reset timers permit a wide range of timing operations in a single unit.

In this activity, a simple reset timer will be used to build a load control circuit that produces either interval or delay timing operations. Initially, when the timer is energized by the control switch, load A is turned on and load B is turned off. After the expired time setting or the "time out" condition has been reached, load A is turned off and load B is turned on. The action of a motor-driven timer is used to represent a type of timing application that might be used in conjunction with robotic system applications.

Equipment and Materials:

- Reset timer (Eagle Signal HD-50 Series) or equivalent.
- 7.5 W incandescent lamps with sockets (2).
- SPST toggle switch.

Procedure:

1. Refer to the timer manufacturer's product manual before attempting to complete this activity.
2. Wire the reset timer for the "on delay" operation of **Figure 5-21**.

Figure 5-21. Reset timer with on delay operation.

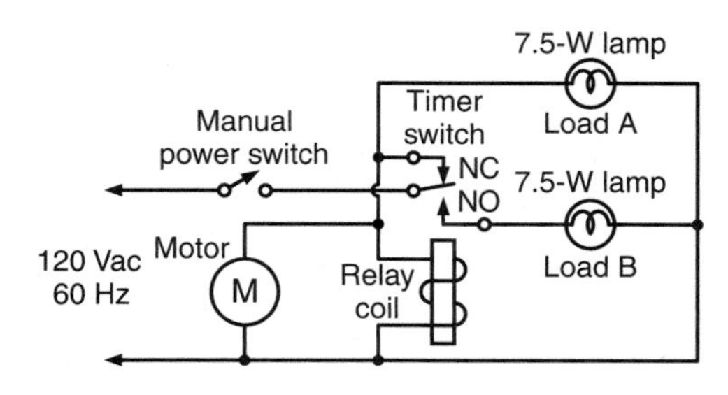

3. Adjust the time setting dial by pulling it out and turning it to a desired setting. Releasing the dial automatically locks it in the new setting.
4. Turn on the manual power switch and describe the operating condition of loads A and B.

 ____

 ____

5. When the dial setting trips, describe the operating condition of the load and the motor.

 ____

 ____

6. Start a new operation cycle by opening the control switch momentarily, then turning it on again. After the cycle has been in operation for a few seconds, momentarily open the control switch, then close it. What action does this initiate?

____

____

Analysis:

1. What are some industrial applications of a reset timer?

____

____

2. Could this timer be modified to achieve a different function of some type? Explain.

____

____

3. Explain the difference between *interval* and *delay* timers.

____

____

Activity 5-12—Digital Timers

Name ______________________ Date ______________ Score ____________

Objectives:

Digital timers employ a time base generator (oscillator), a counter, and a load-driver circuit. The counter circuit frequently has several output terminals that permit selection of time ranges. Control ranges of 1T, 2T, 4T, 8T, 16T, and 32T are typical. Each output is rated in terms of the time base (T) of the generator circuit.

In this activity, you will employ the SE/NE555 timer IC as the time base generator of a simple digital timing system. The output of the generator will be counted and divided to some extent by the range of the system. Through this circuit investigation, you will see and actually work with the basic parts of a simple digital timing system. Commercially designed timers of this type are usually housed in an enclosure that does not permit access to component parts. These timers generally do not employ moving parts and are all solid state. The timing period is very accurate and usually ranges from microseconds to hours. Again, this type of device is being studied to represent the various types of timing control applications that might be used with robotic systems.

Equipment and Materials:

- SE/NE555 IC.
- 200 kΩ, 1/8 W resistors (2).
- 390 Ω, 1/8 W resistor.
- Light-emitting diode.
- 4 μF, 25V dc capacitor.
- 0.01 μF, 100V capacitor.
- 50 μF, 25V capacitor.
- SPST toggle switches or pushbuttons.
- SN7490 IC.
- Circuit mounting board.

Procedure:

1. Construct the experimental digital timer circuit of **Figure 5-22** by first connecting the time base generator. Do not energize the SN7490 at this time.
2. Determine the charging time of the circuit by the $t_c = 0.693\ (R_A + R_B)\ C_1$ formula. Test the accuracy of the circuit. Record your findings.
3. Determine the discharge time of the circuit by the $t_d = 0.693R_B\ C_1$ formula. Test the accuracy of the circuit. Record your findings.

   ____

4. Turn off the power supply and complete the SN7490 counter IC. Connect it to the same power source used by the 555. Connect the second LED indicator to the D output of the 7490 IC. Connect the IC clock source to the output (pin 3) of the 555.
5. Turn on the power source and count the number of timing cycles needed to energize the 7490 LED readout.

   What mathematical function does this represent?

   ____

   ____

Figure 5-22. Digital timer circuit.

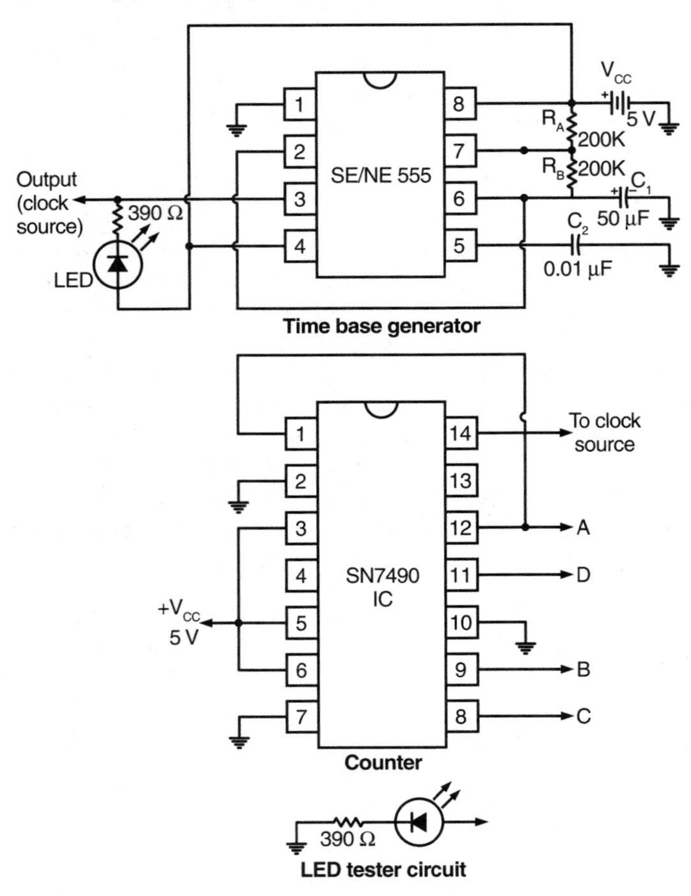

6. Turn off the power source and move the LED readout of the 7490 to output A. Turn on the power again and determine the counting cycles needed to energize the LED.

   ____________________ timing cycles.

   What mathematical function does this represent?

   ____

   ____

7. Turn off the power supply and alter the SN7490 circuitry as indicated in **Figure 5-23**. Change C_1 of the time base generator to 4 μF.

Figure 5-23. Modified counter circuit.

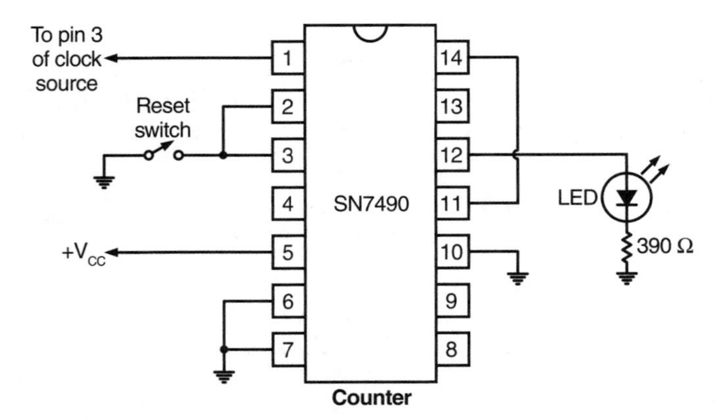

8. Turn on the power supply and open the reset switch. Close the switch during a time when the time base generator LED is off. Then count the number of timing cycles needed to energize the SN7490 LED.

   ________________ timing cycles.

   Then count the number of timing cycles needed to turn off the LED. What does this represent as an output of the IC?

   ____

   ____

Analysis:

1. How could the digital timer of this experiment be extended? Name two ways.

   ____

   ____

2. How can the digital timer of this experiment be made variable?

   ____

   ____

3. Would it be advantageous to have a reset switch an actual digital timer? Why?

   ____

   ____

Activity 6-1—Force through Liquids

Name ______________________________ Date ______________ Score __________

Objectives:

One of the most significant functions of a hydraulic power system is the transmission of force through a liquid. According to Pascal's law, pressure applied to a confined fluid is transmitted equally through the fluid in all directions, and is applied at right angles to the containing surfaces. Fluid does not have a specific shape, which allows it to conform to the internal features of the container. Liquid also has poor compressibility, which means it tends to resist compression when force is applied. These ideas are extremely important characteristics of an operating fluid power system.

In this activity, two hydraulic cylinders are connected to form a simple closed-loop system. Through this system you will be able to observe how force is transmitted through a liquid. You will also have an opportunity to test the compressibility of a hydraulic fluid.

Equipment and Materials:

- Hydraulic fluid or other fluid.
- Double-acting cylinders—1 1/8″ bore (2).
- Assortment of quick-disconnect transparent tubing pieces.
- Safety glasses or goggles.

Safety:

Hydraulic power systems operate at high pressure. Always wear protective eyewear and check all fittings before starting any hydraulic system.

Procedure:

1. Attach a length of flexible tubing to the port opposite the piston end of one of the 1 1/8″ bore double-acting hydraulic cylinders. Position the unit horizontally, with the ram or piston retracted into the cylinder.
2. Place the unattached end of the flexible tubing into a cup of hydraulic fluid. Manually force the piston in and out of the cylinder several times to charge it with oil and clear it of air.
3. Pull the piston all the way out, fully charging it with oil. Set it aside momentarily.
4. Attach a length of flexible tubing to the port nearest the piston end of the second 1 1/8″ bore, double-acting hydraulic cylinder.
5. Using the same procedure outlined in Steps 2 and 3, clear the cylinder of air and charge it with oil. The piston should ultimately be fully retracted when it is charged.
6. Connect the unattached ends of the flexible tubing to the other cylinder as indicated in **Figure 6-1**.
7. Have a lab partner hold cylinder 1 while you hold cylinder 2. Pushing the ram of cylinder 2 down will transfer the force of cylinder 2 through the fluid to cylinder 1.
8. If transparent flexible tubing is used you should be able to see the direction of fluid flow by watching any air bubbles trapped in the system.
9. Alter the position of the two cylinders and repeat the procedure. What do the results indicate about a hydraulic system?

   ____

   ____

Figure 6-1. Hydraulic power test circuit.

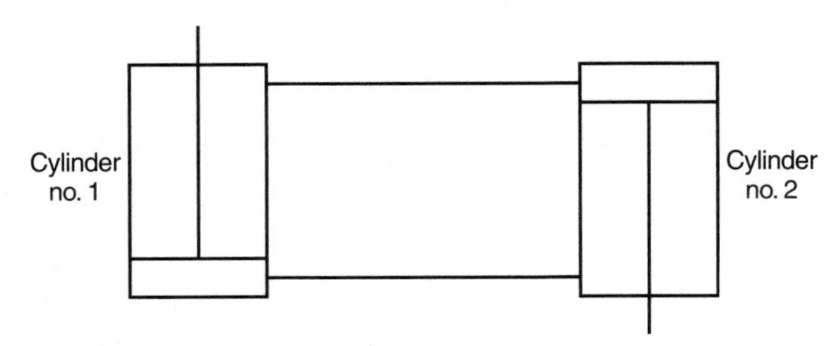

10. Adjust the position of the two pistons so they are near the center of the cylinders. Apply force to both pistons. Ideally there should be no compression of the fluid. Any trapped air in the system will permit some degree of compressibility. How does your system respond?

____

____

11. Momentarily remove the flexible tubing from cylinder 1. Force the cylinder down approximately 1/4″, then pull it back to its original position. Return the flexible tubing to the cylinder.
12. Adjust the position of the two cylinders near the center and repeat Step 10. How does the system respond?

____

____

13. Disconnect the flexible tubing from each cylinder and clear the cylinders of fluid. Wipe excess fluid off all components and return them to the storage area.

Analysis:

1. What is meant by the term *compressibility*?

____

____

2. Explain how force is transferred through a fluid.

____

____

3. Describe the mechanical action of a double-acting cylinder.

____

____

____

____

Activity 6-2—A Static Hydraulic System

Name ______________________________ Date ______________ Score ____________

Objectives:

A hydraulic jack is one of the most common static fluid systems in operation today. With this system, a pump cylinder forces fluid into a ram cylinder through check valves. As a general rule, systems of this type are usually self-contained in a single housing.

In this activity, a simple hydraulic jack system will be constructed with discrete components. Through this activity, you will have an opportunity to observe fluid flow through check valves that will permit an alternate direction of flow.

A hydraulic jack is often described as a static hydraulic system. This refers to the fact that the system develops its own operating pressure when energized by an outside mechanical force. A distinct mechanical advantage is developed by the two cylinders because of the differences in their size.

Equipment and Materials:

- Hydraulic fluid or other fluid.
- Single-acting cylinder—1 1/8″ bore.
- Single-acting cylinder—1/2″ bore.
- Check valves (2).
- Flow control valve (needle type).
- "T" connectors (2).
- Pressure gauge—160 lb/in^2.
- Assortment of quick-disconnect transparent tubing pieces.
- Safety glasses or goggles.

Procedure:

1. Assemble the hydraulic circuit of **Figure 6-2**. The 1 1/8″ bore cylinder will serve as the ram of the hydraulic jack, and the 1/2″ bore cylinder as the pump. Mount both cylinders in an upright position.

Figure 6-2. Hydraulic jack circuit.

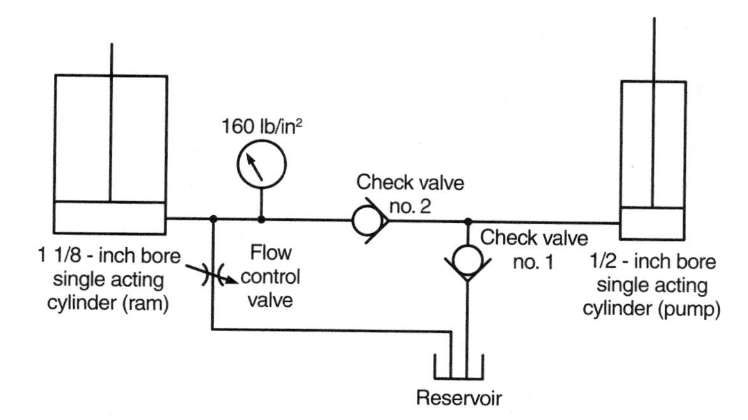

2. Place the rams of both cylinders in the retracted position before attaching the flexible tubing to the cylinders, and close the flow control valve. Record the stoke length of the ram cylinder ______________ and the pump cylinder ______________.

3. Pull the ram of the pump cylinder out of the cylinder, then down. One downward stroke of the pump cylinder causes the ram cylinder to move __________________ millimeters/inches.
4. How many strokes of the pump are needed to fully extend the ram?

   __________________ strokes.
5. What pressure is indicated on the gauge?

   Pressure = __________________ psi (lb/in^2).
6. Open the flow control valve to drain the ram cylinder. You will need to slowly force the ram to its retracted position. Then close the valve to operate the system again.
7. Attach a weight of approximately ten pounds to the ram cylinder and repeat the operation of the jack.
8. What noticeable changes occur when the system is forced to do work?

   ____

   ____
9. Disconnect the system, drain the fluid from the cylinders and the tubing. Wipe any excess fluid from the components and return them to the storage area.

Analysis:

1. How much work was done by the ram cylinder in Step 7?

   ____
2. Describe the fluid circulation action that takes place during the operation of this system.

   ____

   ____
3. Does the pressure gauge change during the pumping action?

   ____
4. Why does this system have a mechanical advantage?

   ____

   ____

   ____

Activity 6-3—A Hydraulic Ram System

Name ______________________________ Date ______________ Score __________

Objectives:

Hydraulic presses or rams play a very important role in industry today. Mechanical systems of this type are actuated by an electric-motor-driven pump that serves as the source of hydraulic energy. The transmission path of the system distributes energy to the load device, which does some form of mechanical work. Flow and directional control are achieved by special valves in the transmission line. An indicator is used to display system pressure. The hydraulic fluid is sent back to the pump reservoir through a return line. Systems of this type are classified as active units when compared with the static systems of Activities 6-1 and 6-2.

In this activity, you will have an opportunity to identify some basic hydraulic components and their associated diagram symbols. You will then assemble these components into a functioning hydraulic ram system. You will be able to observe the change in the direction of fluid flow by valve action. The basic operating principle of this hydraulic system has widespread application in industrial machinery today.

Equipment and Materials:

- Hydraulic fluid or other fluid.
- Electric hydraulic power unit.
- Flow control valve.
- Two-position, four-way, lever-action control valve.
- Double-acting cylinder—1 1/8″ bore.
- Pressure gauge—160 lb/in^2.
- "T" connector.
- Assortment of quick-disconnect transparent tubing pieces.
- Safety glasses or goggles.

Procedure:

1. Select the necessary components to construct the hydraulic ram circuit of **Figure 6-3**.

Figure 6-3. Hydraulic ram circuit.

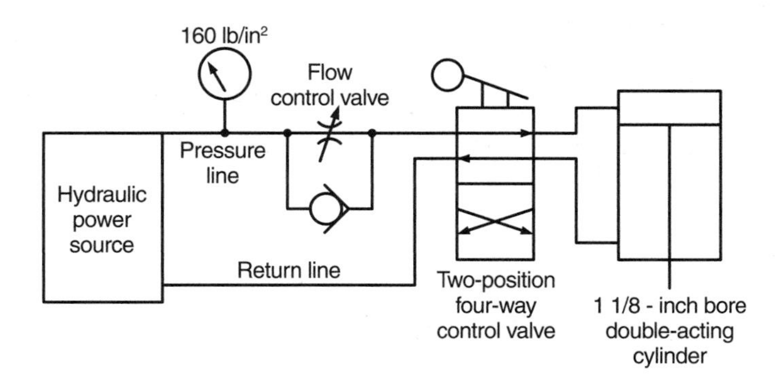

2. Start the electric pump motor of the hydraulic power source. Slowly open the shutoff control valve of the circuit while observing the pressure gauge. Depending on the initial position of the four-way control valve, the ram of the cylinder may extend itself slowly. Record the system pressure.

   Pressure = ______________ psi (lb/in^2).

3. With the four-way control valve positioned as indicated in the circuit diagram, move the handle to the other position. What influence does this control action have on the ram of the double-acting cylinder?

____

____

4. Move the handle of the four-way control valve back to its original position to see the control action on the cylinder. Describe the action of the cylinder's ram.

____

____

5. The direction of fluid flow can be readily traced by the movement of air bubbles in the transparent tubing. Switch the control valve to extend the ram and observe the direction of fluid flow. If the bubbles are not present, they can be introduced by closing the flow control valve and momentarily disconnecting the return line from the four-way control valve. Depress the check valve in the line and let a small amount of hydraulic fluid drain from the line. Then return the line to the control valve and turn on the flow control valve.
6. Indicate the direction of fluid flow on the circuit diagram with solid line arrows.
7. Switch the control valve to a position that will cause the ram of the cylinder to retract. Trace the fluid flow through the circuit. Make a sketch of the flow direction with broken line arrows or with a different colored pen or pencil on the circuit diagram.
8. Turn off the electric pump motor of the hydraulic power source and disconnect the circuit. Wipe off any excess hydraulic fluid from the components and return them to the storage area.

Analysis:

1. What are the primary system functions of the circuit constructed in this activity?

____

____

2. What specific component of the circuit is used to achieve each system function?

____

____

3. How could the ram in this circuit be used in an industrial application?

____

____

4. How does this system differ from the static systems of Activities 6-1 and 6-2?

____

____

Activity 6-4—Hydraulic Power Source

Name ______________________ Date ______________ Score ____________

Objectives:

The power source of a hydraulic system is a pump. This component is responsible for producing a fluid flow that will ultimately develop pressure as it encounters resistance as it passes through system components. A hydraulic pump must be in continuous operation whenever the system is operational. Through its action, hydraulic fluid is forced to move at a specific flow rate, which causes a corresponding pressure to occur when resistance is encountered.

In this activity, you will study the basic components of a hydraulic power unit. This will include the hydraulic pump, reservoir, filters, pressure relief valve, pressure manifold, and return manifold. Pressure developed by this power source is variable, with a maximum rating of approximately 300 psi (lb/in^2). For most experimental applications, the power source needs values of 100 lb/in^2 or less. Typical industrial hydraulic power sources have values in the 1000 to 3000 psi (lb/in^2) range.

Equipment and Materials:

- Hydraulic fluid or other fluid.
- Hydraulic power source.
- Pressure gauge—160 lb/in^2.
- Assortment of quick-disconnect transparent tubing pieces.
- Measuring container—1/2 gal.
- Open-ended disconnect tubing piece.
- Safety glasses or goggles.

Procedure:

Section A. Power Source Analysis

1. A basic hydraulic power source is needed to perform this activity. As a general rule, most of the components of this unit are permanently mounted and ready for operation.
2. Locate the electric motor of the unit and plug it into the appropriate electrical outlet. Trace out the electrical circuit and locate the motor/pump on/off switch. In many units, the pump and motor may be mounted in a single housing.
3. Locate the system reservoir. Trace the fluid flow path from the reservoir to the inlet side of the hydraulic pump. Does your unit employ a fluid strainer or filter in this line?

   ____

4. Trace the fluid path from the outlet side of the pump to the pressure manifold. Are there any filters or controls in this part of the system?

   ____

5. Where is the pressure relief valve of your system located?

   ____

6. Turn on the electric motor. Observe the pressure gauge and record the system pressure. If at all possible, see at what pressure the system returns fluid to the reservoir. This is determined by the pressure relief valve setting.
7. Connect a length of tubing to one of the pressure manifold valves. Run the other end of the tubing to an open tray or to the reservoir return line. Open the manifold valve momentarily until all of the air bubbles pass from the line. Does the pressure gauge change during this operation?

   ____

8. Turn off the electrical power to the pump drive motor. In the following space, make a sketch of your hydraulic power source pointing out specific component locations:

9. In the following space, draw a schematic diagram of the hydraulic power source using appropriate fluid power symbols:

Section B. Hydraulic Pump Delivery Rate

1. Attach a 160 lb/in^2 gauge to one of the pressure manifolds of the hydraulic power source. Connect a length of plastic tubing to the output side of a second manifold valve. Then connect the other end of the tubing to one of the return manifold connections.
2. Ready the hydraulic power source for operation by applying electricity and turning on the pump motor. Open the pressure manifold valve attached to the pressure gauge. Record the system pressure. This valve also indicates the pressure setting or value of the pressure relief valve. If more than 100 lb/in^2 is indicated, ask the instructor to show you how to adjust the pressure limit.
3. Gradually open the pressure manifold valve with the tubing attached. How does the system pressure gauge respond as this valve is opened?

 ____

4. Close the pressure manifold valve and remove the plastic tubing with quick disconnects on each end. Replace it with a piece of tubing with a disconnect on only one end. The open end of the tubing will be used to fill a half-gallon measuring container to determine the pump delivery rate.
5. Open the manifold valve enough to maintain 100 lb/in^2 with the full output flowing through the tube. Fill the container and record the time it takes to pump 1/2 gallon. Run two trial tests at this pressure and average your results. Record your data in the chart in **Figure 6-4**.

Figure 6-4. Hydraulic pump delivery data.

Pressure (lb/in^2)	Half-gallon filling time (seconds)	Pump delivery rate (GPM)	Calculated horsepower*
100			
80			

$$\text{*Horsepower} = \frac{\text{gal/min} \times \text{lb/in}^2}{1714}$$

Name ______________________________

6. Repeat the procedure outlined in Step 5 for a pressure setting of 80 lb/in^2.
7. Calculate the horsepower delivered by the pump for each pressure setting, and record the data in the chart.
8. Turn off the power unit and disconnect all circuit components. Wipe them free of excess fluid and return them to the storage area.

Analysis:

1. Explain the operation of the hydraulic power unit investigated in this activity.

____

____

2. Explain how the system pressure is altered.

____

____

3. What is the function of the pressure relief valve?

____

____

4. What type of displacement does the pump of the hydraulic power unit supply?

____

Activity 6-5—Hydraulic Pumps

Name ______________________________ Date ______________ Score __________

Objectives:

A hydraulic pump is designed to provide an appropriate fluid flow that will ultimately develop system pressure (based on the encountered resistance to flow). In an operating system, the pump accepts power from a drive motor (prime mover) and converts it into fluid flow.

Fluid is pulled into the inlet port of the pump and is expelled (forced out) from the outlet port. After passing through the pump, fluid flow occurs at a faster rate. Ultimately, this increase in fluid flow causes an increase in system pressure when the flow encounters component resistance. Most hydraulic pumps displace a certain amount of fluid (volumetric displacement) with each revolution.

The gear pump in this activity is used to change rotary motion into fluid energy. Essentially, this pump contains two gears enclosed in a precision machined housing. Rotary motion from the power source is applied to the drive gear. Rotation of this gear causes the second, or driven gear to turn, with their teeth meshing in the middle. The unmeshed gear teeth carry fluid from the inlet port to the center where the teeth mesh together and force the fluid to pass from the outlet port.

In this activity, you will construct a simple gear pump circuit and rotate the shaft of the pump manually to observe the direction of fluid flow. This will provide you with an opportunity to see how pump rotation and shaft speed influence the output of a hydraulic pump.

Equipment and Materials:

- Hydraulic fluid or other fluid.
- Bidirectional, fixed-displacement, gear motor/pump.
- Flow control valve.
- Double-acting cylinder—1 1/8″ bore.
- Assortment of transparent flexible tubing with disconnects.
- Safety glasses or goggles.

Procedure:

1. Connect the simple hydraulic pump circuit in **Figure 6-5**. The gear motor in this case will be used as a pump.

Figure 6-5. Hydraulic pump circuit.

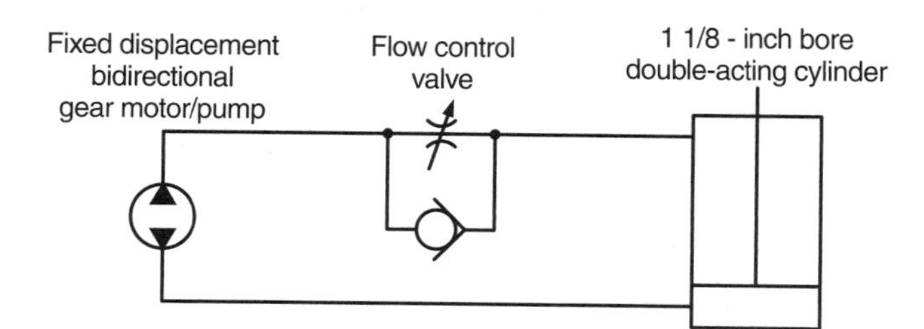

2. Rotate the shaft of the pump in a clockwise direction with one hand while holding it with the other hand. The tubing pieces should contain enough fluid to make the system operational.
3. Continue to rotate the shaft until the ram of the double-acting cylinder begins to move.
4. Air trapped in the tubing pieces can be used to trace direction of fluid flow. Clockwise pump rotation causes the hydraulic fluid to flow in a ________________ direction.

5. Rotate the shaft until the cylinder ram stops moving.
6. Rotate the shaft in a counterclockwise direction. Counterclockwise pump rotation causes hydraulic fluid to flow in a ____________________ direction.
7. Disconnect the circuit components and wipe off all excess fluid with a cloth. Return all components to the storage area.

Analysis:

1. How does pump speed of rotation influence the flow rate?

   ____

2. How does a gear pump produce fluid flow?

   ____

   ____

3. Explain the basic principle of fluid flow as achieved by a gear pump.

   ____

   ____

4. Explain how the volumetric displacement of the pump is affected with changes in the pumps rotational speed.

   ____

   ____

Activity 6-6—Pressure Control Valves

Name ______________________ Date ______________ Score ____________

Objectives:

When a hydraulic pump is in operation, it supplies a defined amount of fluid to the system with each revolution of the drive motor. With the hydraulic pump operating continuously, there is a need for an alternate flow path for fluid not used by the system. Pressure relief valves placed in the power source are designed to provide an alternate flow path when the need arises.

A pressure relief valve is a type of check valve that permits flow only when a certain pressure is reached. At reduced pressure, the valve checks the return flow and permits the fluid to pass into the system. At higher pressure levels it passes fluid back into the reservoir. In this application, the relief valve is a safety device.

In this activity, you will have an opportunity to investigate the action of both a pressure relief valve and check valves more carefully. You will also have an opportunity to test the "cracking" pressure of the pressure relief valve. This refers to the release pressure that causes flow into the reservoir.

Equipment and Materials:

- Hydraulic fluid or other fluid.
- Hydraulic power source.
- Pressure relief valve.
- Flow meter.
- Flow control valve.
- Assortment of quick-disconnect hoses.
- Pressure gauge—160 lb/in^2.
- Safety glasses or goggles.

Safety Note:

You will be dealing with high system pressures in this activity. Follow all applicable safety procedures. Do not allow system pressure to become excessively high.

Procedure:

1. Select the components needed to assemble the circuit of **Figure 6-6**. Notice the two paths for the fluid flow in this circuit. This represents a typical power source system divider.

Figure 6-6. Hydraulic power source system divider circuit.

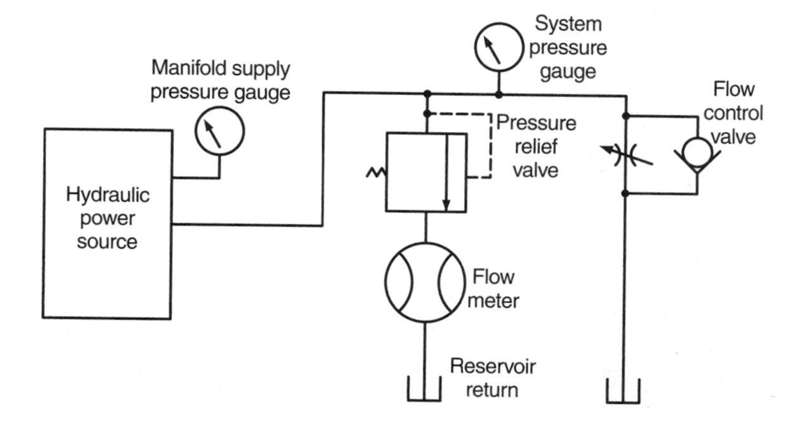

2. Adjust the pressure relief valve in the counterclockwise direction until it stops, then back off one turn of the valve control. Close the flow control valve completely.
3. Start the power unit and adjust the pressure relief valve to read 100 lb/in^2 of system pressure.
4. Open the flow control valve completely while observing system pressure. How does the pressure respond? Pressure with the flow control valve open is ________________ lb/in^2.
5. Close the flow control valve again while observing the system pressure gauge. What does the flow meter do when the flow control valve is turned off?

   ____

6. Adjust the flow control valve to produce 100 lb/in^2 of system pressure. Then note the exact reading where flow first occurs through the flow meter. This is the cracking pressure of the valve.
7. Adjust the pressure relief valve one more turn as instructed in Step 2. Then adjust the flow control valve to 150 lb/in^2 of system pressure and determine the new cracking pressure of the pressure relief valve.
8. Turn off the power source and disconnect the circuit. Wipe all excess oil from the components and return them to the storage area.

Analysis:

1. What is the function of the pressure relief valve in a system of this type?

   ____

   ____

2. Define the term *cracking pressure.*

   ____

   ____

3. What is the function of a flow control valve in a hydraulic circuit?

   ____

   ____

4. What accounts for a difference in manifold supply pressure and system pressure?

   ____

   ____

   ____

   ____

Activity 6-7—Directional Control Valves

Name ______________________________ Date ____________ Score __________

Objectives:

A directional control valve (DCV) of a hydraulic system is designed to alter the fluid flow in such a way that it can start, stop, or reverse the flow of the fluid to the system components. Valves that achieve this type of control are described as two-, three-, and four-way directional control valves with different position settings. Controls of this type may be actuated manually, with fluid pressure, with electricity, or by mechanical energy.

In this activity, you will construct a simple hydraulic motor control circuit and test the direction of flow through the circuit with a flow meter. As a general rule, this type of flow indicator will only respond to flow in a single direction. You will also be able to test the direction of flow with respect to different circuit positions in order to see how this device alters the fluid flow. Through this experience, you will become more familiar with the basic components of a hydraulic system and be able to trace out the flow path from the manifold source to the reservoir.

Equipment and Materials:

- Hydraulic fluid or other fluid.
- Hydraulic power source.
- Pressure relief valve.
- Flow meter.
- Three-position, four-way directional control valve.
- Hydraulic gear motor.
- Assortment of quick-disconnect hoses.
- Safety glasses or goggles.

Procedure:

1. Select the components needed to construct the hydraulic circuit of **Figure 6-7**. Note the location of the four test points. In this activity, you will be asked to insert the flow meter into each test point and observe the direction of flow. The flow meter will only respond when the flow is from the inlet port to the outlet port. Flow in the reverse direction will not produce an indication on the meter.

Figure 6-7. Hydraulic motor directional control circuit.

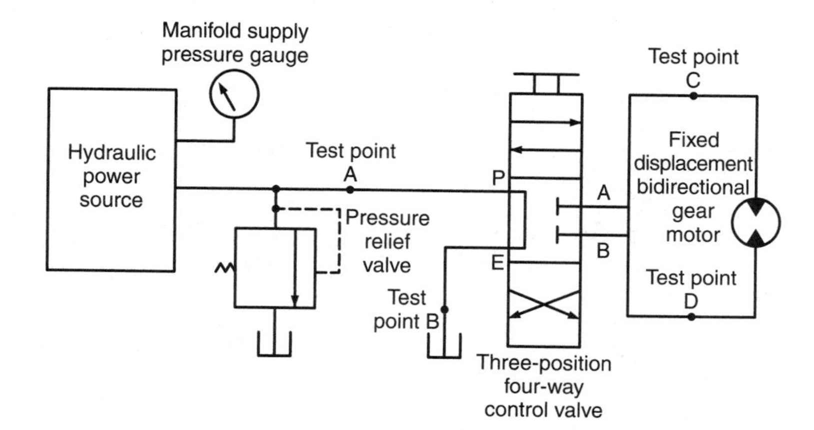

2. Close the pressure relief valve and back it off one complete turn. Then turn on the hydraulic power source and adjust the pressure relief valve for a manifold pressure of 100 lb/in^2 as indicated on the manifold supply gauge.
3. Momentarily turn off the power source and insert the flow meter into test point A. Then turn the power source on again.
4. Pull the lever of the four-way valve to the fully extended position, which should cause rotation of the motor. Record the response of the flow meter and the motor.
5. Shift the directional control valve to the center position and record the flow meter and motor response.
6. Push the lever to the retracted position and note the flow meter and motor response.
7. Move the flow meter to test points B, C, and D, and repeat Steps 4, 5, and 6. Record the direction of flow and motor response for each situation.
8. In the following space, draw three separate schematic diagrams with arrows showing the direction of fluid flow for each position of the four-way valve.

9. Turn off the hydraulic power source and disconnect the circuit. Wipe all components free of excess fluid and return them to the storage area.

Analysis:

1. What do the letter designations P and E on the four-way valve denote?

____

2. What are the three motor operating conditions of the four-way valve?

____

____

3. Would the four-way valve be classified as a full-control or partial-control device?

____

____

4. How does the way the hoses are connected from the DCV (ports A and B) affect the motor operation?

____

____

Activity 6-8—Hydraulic Linear Actuators

Name ______________________________ Date ______________ Score ___________

Objectives:

The fundamental purpose of a hydraulic power system is to produce some form of mechanical work. The part of the system responsible for achieving this function is called the load. The term *load,* however, is a rather broad term used to describe all components that consume power and do work. The term *actuator* is used more commonly to describe specific components in hydraulic systems that achieve the work function.

Linear actuators are designed to lift, hold, compress, or clamp objects during different manufacturing processes. Hydraulic cylinders are the most common linear actuator in operation today. These devices come in a variety of shapes and sizes that range from a fraction of an inch to several feet in diameter and length. The amount of force a cylinder can produce is determined by the surface area of the piston inside it and the pressure applied to it. Therefore, a cylinder with a larger surface area can produce more force than a smaller cylinder with the same pressure applied. Both single-acting and double-acting cylinders are included in the linear actuator classification.

In this activity, you will have an opportunity to construct a simple series and parallel linear actuator circuit and test the pressure and circuit operation under load and no load conditions. Through this experience you will become familiar with different system circuit configurations and observe some of the basic principles of series and parallel connected systems. Each type of circuit has a unique set of characteristics with respect to system pressure.

Equipment and Materials:

- Hydraulic fluid or other fluid.
- Hydraulic power source.
- Pressure relief valve.
- Three-position, four-way control valve.
- Pressure gauge—160 lb/in^2.
- Double-acting cylinders—1 1/8″ bore (2).
- "T" hose connectors (3).
- Assortment of quick-disconnect hoses.
- Safety glasses or goggles.

Procedure:

Section A. Series Linear Actuators

1. Select the appropriate components needed to construct the hydraulic circuit of **Figure 6-8**. Either a cylinder loading device or an actual weighted load needs to be attached to the cylinder in some steps of this activity.
2. Turn on the power source and adjust the pressure relief valve to 120 lb/in^2.
3. Push the control lever of the four-way valve into the valve while observing the pressure at test point A. Make this reading while the cylinders are moving.

   Test point A pressure = ________________ lb/in^2.
4. Move the four-way valve to the center or "off" position and move the pressure gauge to test point B. Repeat Step 3 and record the pressure.

   Test point B pressure = ________________ lb/in^2.

Figure 6-8. Hydraulic linear actuator circuit—actuator in series

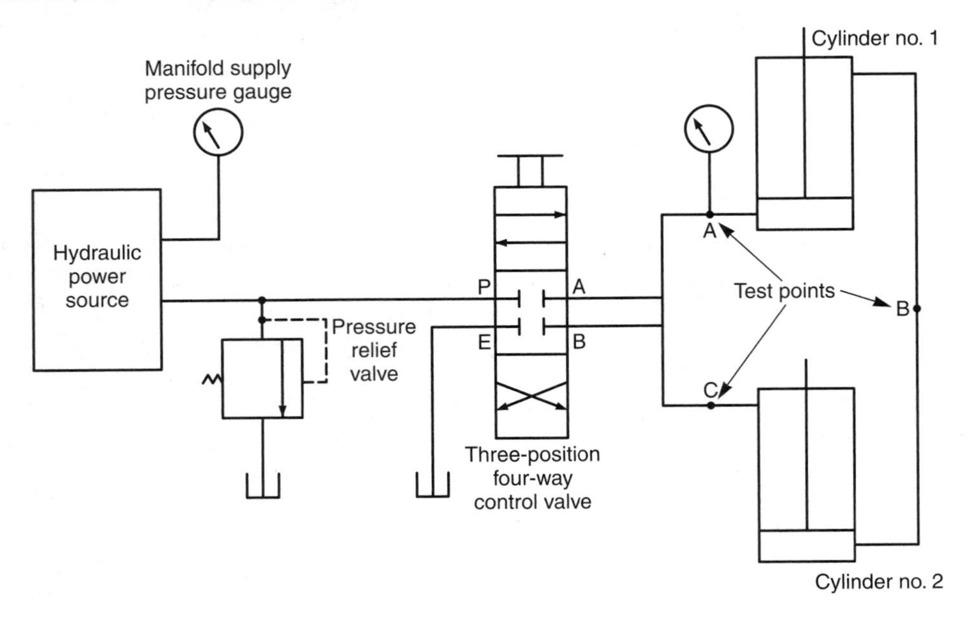

5. Return the four-way valve to the center or "off" position and move the pressure gauge to test point C. Repeat Step 3 and record the pressure.

   Test point C pressure = ________________ lb/in^2.

6. The pressure tests you just performed represent the no-load test condition. If a cylinder loading device is available, connect it to cylinder 1 and adjust it to the heavy load condition. If a cylinder loading device is not available, attach a small platform plate to cylinder 1. Place 20 pounds of weight on the platform with the cylinder mounted in the vertical position.

7. Repeat Steps 3, 4, and 5 with the cylinder loaded. The pressure at test point C should be taken first since the gauge should be at this location from the previous test.

8. Make a chart showing the relationship of pressure between load and no-load conditions at each test point.

9. Increase the system pressure to 160 lb/in^2 and test the operation of the circuit in its loaded condition. How does this added pressure alter cylinder operation?

   ____

10. Reduce the system pressure to 80 lb/in^2 and test the operation of the circuit in its loaded condition. How does it respond to reduced pressure?

    ____

11. Turn the power source off and disconnect the circuit.

Section B. Parallel Linear Actuators

1. Select the components needed to construct the hydraulic circuit of **Figure 6-9**.

2. Turn on the power source and adjust the pressure relief valve to 120 lb/in^2.

3. Push the control lever of the four-way valve into the valve while observing the pressure at test point A.

   Test point A pressure = ________________ lb/in^2.

   Now pull the control lever completely out while observing the pressure. Make a pressure reading while the cylinders are in motion.

   Pressure = ________________ lb/in^2.

   Does the flow direction influence the pressure?

   ____

Name ____

Figure 6-9. Hydraulic linear actuator circuit—actuator in parallel.

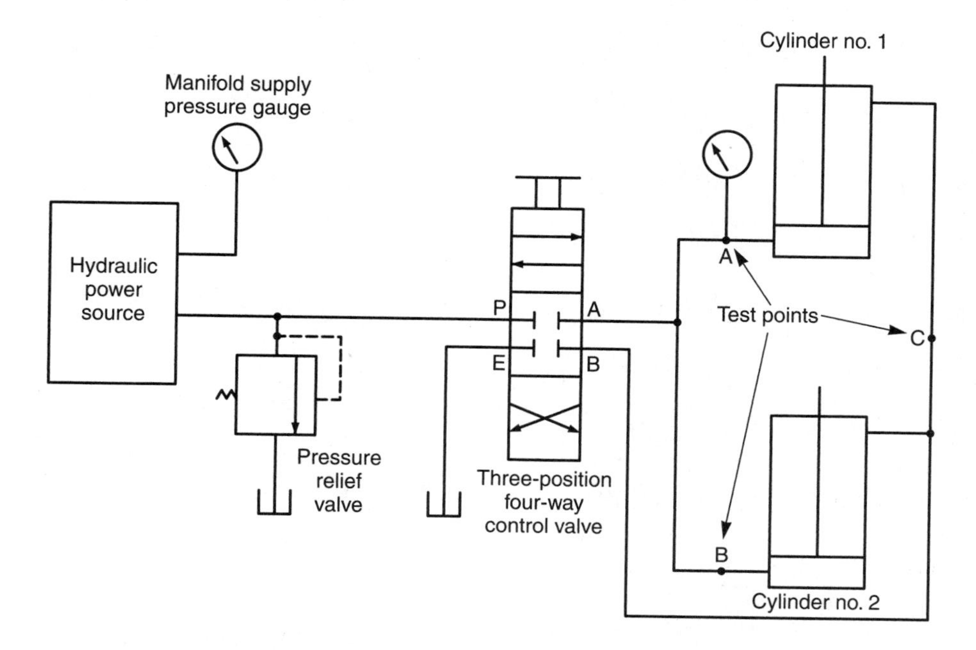

4. Place the four-way control valve in the center or "off" position. Place the pressure gauge at test points B and C and repeat the procedure outlined in Step 3.

   Test point B pressure = ____________________ lb/in^2.

   Test point C pressure = ____________________ lb/in^2.

5. These tests represent the no load condition of the linear actuator. If a cylinder loading device is available, connect it to cylinder 1 and adjust it to the heavy load condition. If this device is not available, attach a small platform plate to cylinder 1. Place 20 pounds of weight on the platform with the cylinder mounted in the vertical position.

6. Repeat Steps 3, 4, and 5 with the cylinder loaded. The pressure at point C should be taken first since the gauge should be at this location from the previous test.

7. Make a chart showing the relationship between load and no-load pressure conditions at each test point.

8. Increase the system pressure to 160 lb/in^2 and test the operation of the circuit in its loaded condition. How does this added pressure alter the operation of the cylinders?

   ____

9. Reduce the system pressure to 70 lb/in^2 and test the operation of the circuit in its loaded condition. How does it respond to reduced pressure?

____

10. Turn off the power source and disconnect the circuit. Wipe all components free of excess fluid and return them to the storage area.

Analysis:

1. Why is this circuit described as a series linear actuator?

____

____

2. What influence does loading have on the pressure of a series linear actuator circuit?

____

3. What is a characteristic of the series circuit with respect to pressure throughout the system?

____

4. What general influence does system pressure have on the operational speed of a series linear actuator circuit?

____

5. Why is this circuit classified as a parallel linear actuator?

____

6. What influence does loading have on the system pressure of a parallel linear actuator system?

____

____

7. What is a characteristic of the parallel circuit with respect to pressure throughout the system?

____

____

8. What influence does system pressure have on the operational speed of a linear actuator?

____

____

Activity 6-9—Hydraulic Rotary Actuators

Name ______________________ Date ______________ Score __________

Objectives:

Hydraulic rotary actuators are designed to produce an appropriate form of rotary motion in either direction. Fluid under pressure flows through the rotating member and causes it to move before it is expelled from the outlet side. Counterclockwise rotation is achieved by reversing the direction of fluid flow. These devices are used in industry today for opening and closing gates, for indexing operations, and for lifting operations. The speed of hydraulic rotary actuators is rather slow as compared to electric actuators, but they develop a great deal of torque.

In this activity, a hydraulic gear motor/pump device will be used as a rotary actuator. The turning ability of this device is a measure of the work the device is designed to do. Torque is used to describe the developed amount of work. It is determined by the developed force multiplied by the radius of the rotating arm. You will also have an opportunity to observe the influence of pressure and loading on the rotational speed of the actuator. Rotary actuators represent a very important hydraulic system load device.

Equipment and Materials:

- Hydraulic fluid or other fluid.
- Hydraulic power source.
- Pressure relief valve.
- Three-position, four-way control valve.
- Flow meter.
- Pressure gauge—160 lb/in^2.
- Bidirectional, fixed-displacement, gear motor/pump.
- "T" connector.
- Assortment of quick-disconnect hoses.
- Tachometer.
- Safety glasses or goggles.

Procedure:

1. Select the appropriate components needed to construct the hydraulic circuit of **Figure 6-10**.

Figure 6-10. Hydraulic rotary actuator circuit.

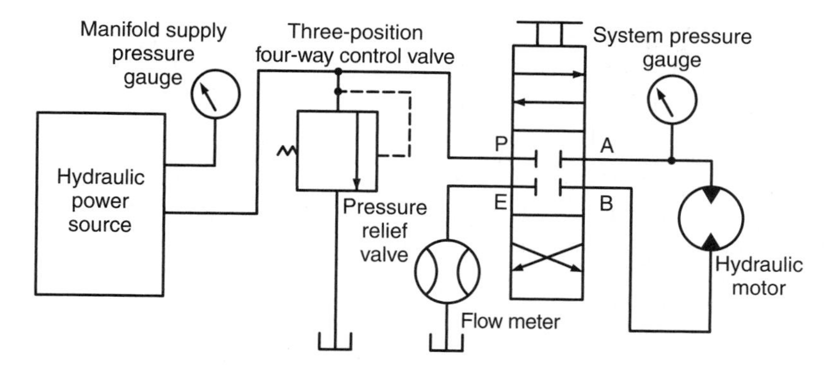

2. Turn on the power source and adjust the pressure relief valve for 120 lb/in^2.
3. Push the control lever of the four-way valve into the valve while observing the direction of motor rotation.

   Direction of rotation is ________________ (clockwise or counterclockwise).

4. Pull the valve control lever to its most extended position.

   Direction of rotation is __________.

5. Perform this test again and note the system pressure needed to produce rotation.
6. The flow meter indicates __________________ gallons per minute (gpm).
7. If a tachometer is available, measure and record the rotational speed of the motor. You can also measure speed when it is slow enough by marking the shaft and counting the number of revolutions per minute.

   Rotational speed = __________________ rpm.

8. This represents the unloaded speed and pressure of the motor. Increase the pressure to 160 lb/in^2. Record the influence this has on speed and flow rate.

   Speed = __________________ rpm.

   Flow rate = __________________ gpm.

9. Reduce the pressure to 80 lb/in^2 and record the influence this has on speed and flow rate.

   Speed = __________________ rpm.

   Flow rate = __________________ gpm.

10. Attach a rotary load of some type to the mounted hydraulic motor.
11. Repeat Steps 2–10 with a substantial load on the motor. Record your findings in the chart in **Figure 6-11**.

Figure 6-11. Hydraulic rotary flow rate.

Pressure (lb/in^2)	Speed (RPM)	Flow rate (GPM)
80		
120		
160		

12. Turn off the power source and disconnect the circuit. Wipe all components free of excess hydraulic fluid and return them to the storage area.

Analysis:

1. What influence does system pressure have on motor speed and flow rate?

   ____

   ____

2. What influence does loading have on pressure and speed?

   ____

   ____

3. Explain how a hydraulic motor produces rotation.

   ____

   ____

4. What does the term *displacement* refer to in hydraulic motors?

   ____

   ____

Activity 6-10—Compressibility of Air

Name ______________________ Date ______________ Score ___________

Objectives:

The most significant difference in a hydraulic and a pneumatic power system is the compressibility of air with respect to fluid. Pneumatic systems are designed to compress air into smaller volumes and increase its pressure. Compressed air can then be stored in a receiver tank and released when the system demands it. This removes the necessity of continuous pump operation that is characteristic of hydraulic systems.

The pressure associated with a pneumatic system can be expressed in a number of ways. Barometric, or absolute pressure, is 14.7 lb/in^2 at sea level. Gauge pressure is the difference in absolute pressure and an unknown system pressure. A gauge reading of 100 lb/in^2 represents an absolute pressure of 114.7 lb/in^2. Pressure below atmospheric pressure is commonly considered as a negative pressure or vacuum.

When a certain volume of air or gas is compressed, it causes a corresponding increase in pressure. Essentially, a decrease in container volume causes a corresponding increase in the density of the molecules. This, in turn, increases the rate at which the molecules strike the walls of the container, which causes a corresponding increase in pressure. Likewise, when the volume of a container is increased, it causes the gas molecules to spread apart. This in turn causes a corresponding reduction in total container pressure. This relationship is normally expressed by the ratio of: $P_1 \times V_1 = P_2 \times V_2$. P_1 and V_1 refer to the pressure and volume of the original container, while P_2 and V_2 refer to its reduced or expanded condition.

In this activity, you will have an opportunity to construct a simple pneumatic system and test the compressibility of air while observing the changes in pressure. The internal, or blind-end, volume of a cylinder is based on the area of the piston times the length of the stroke. Cylinder area is measured in square inches or square centimeters and is determined by the formula

Area = $D^2/4$.

The blind-end volume of a cylinder is determined by the cylinder area multiplied by the stroke length, and is expressed in cubic inches or cubic centimeters.

Equipment and Materials:

- Air supply source—0–100 lb/in^2.
- Check valve.
- System pressure gauge—100 lb/in^2.
- Double-acting pneumatic cylinder—1 1/8″ bore.
- Assortment of quick-disconnect hoses.
- Safety glasses or goggles.

Safety Note:

Pneumatic systems, like hydraulic systems, operate at high pressure. Care should be exercised while working with pneumatic systems. Follow all safety procedures and use protective eyewear.

Procedure:

1. Construct the pneumatic circuit of **Figure 6-12**. Note that this circuit includes several items that are part of the primary air supply. These items are optional and may not be included in the air supply feeding each work area. The part of the circuit that you are primarily concerned with begins at the manifold supply valve.

2. Close manifold supply valves 1, 2, and 3 and open the manual shutoff valve of the air source. Adjust the system pressure regulator to read 15 lb/in^2 on the system pressure gauge.

Figure 6-12. Pneumatic power test circuit.

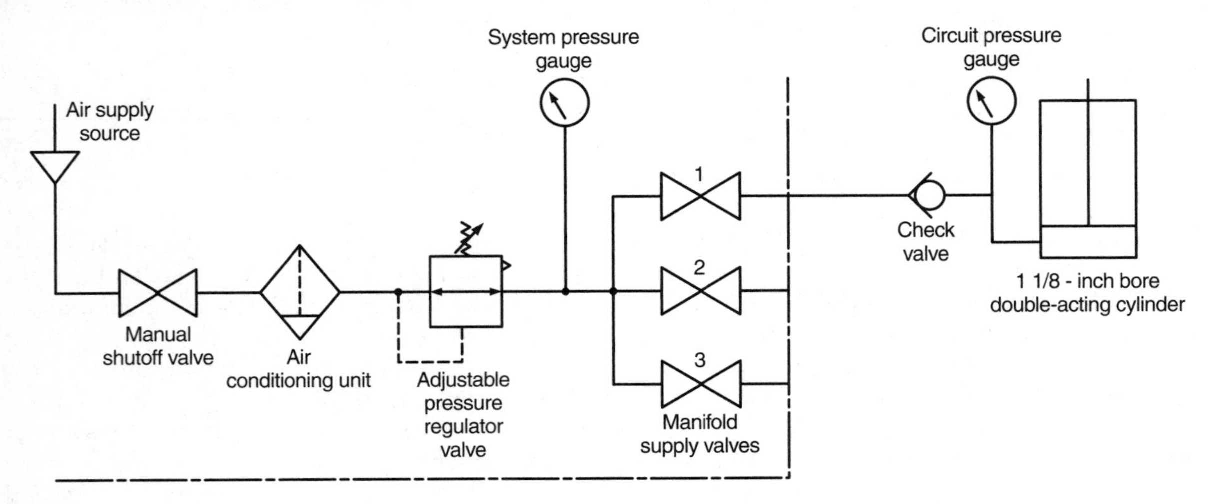

3. Measure the length of the cylinder shaft when it is fully retracted.

   Length = _________________ inches.

4. Slowly open manifold supply valve 1 while observing the action of the gauge and the cylinder. If the circuit does not respond, turn off the manifold valve and check the flow direction of the check valve.

5. When the cylinder ram is fully extended, measure and record the circuit pressure.

   Pressure = _________________ lb/in^2.

   Measure and record the extended length of the cylinder ram.

   Extended length = _________________.

   Calculate the blind-end volume of the cylinder.

   Cylinder volume = _________________.

6. Manually force the piston back into the cylinder one-half of its extended distance while observing the circuit pressure gauge.

   Pressure = _________________ lb/in^2.

   Measure the extended rod length and calculate its blind-end volume.

7. Release the cylinder piston and let it extend again.

8. Adjust the supply regulator to 25 lb/in^2 and repeat Steps 3, 4, 5, and 6. How does this action differ from the preceding steps?

   ____

   ____

9. Close the manifold supply valve and the main air supply shutoff valve. Disconnect the circuit and return all components to the storage area.

Analysis:

1. Describe the relationship between blind-end cylinder volume and pressure.

   ____

   ____

2. What is the absolute pressure of Step 5?

   ____

Name ______________________________

3. As pressure increases how does its compressibility change?

____

4. Why does pneumatic pressure increase when the volume of a container is reduced?

____

____

Activity 6-11—Pneumatic Source Analysis

Name ______________________________ Date ______________ Score __________

Objectives:

The air supply source is an extremely important part of a pneumatic system. It is usually energized by an electric motor that drives a compressor. The compressor, as the name implies, compresses air and forces it into the receiving tank. The receiving tank then holds the compressed air until needed by the system. Depending on the demand for air, the compressor's electric drive motor may only be called upon to operate when the air supply diminishes. The basic principle of air compression and storage represents a unique characteristic of a functional pneumatic system. Basic compressor types include reciprocating piston, centrifugal, and screw type.

In this activity, you will have an opportunity to study the supply system that feeds air to your workstation. This part of the system typically includes a motor-driven compressor, a receiver tank, air conditioning equipment, pressure control valves, regulators, a distribution manifold, and pressure limit switches that respond to tank pressure changes. The air pressure capacity of the compressor is designed to meet the demands of the entire system. Typical industrial pneumatic systems operate within a range of 80 to 200 lb/in^2 gauge pressure.

Equipment and Materials:

- Operating air supply source.

Procedure:

1. With the assistance of the instructor, locate the primary source of the pneumatic system supplying your workstation.
2. Describe the basic type of compressor employed by the system.

   ____

   ____

3. What are the essential parts of this specific air source?

   ____

   ____

4. Make a symbol diagram of the complete air source.

5. What is the maximum receiver tank pressure developed by the system?

   ____

6. At what pressure does the compressor motor turn on to build up receiver tank pressure?

   ____

7. Describe the distribution path from the receiver tank to your workstation.

   ____

   ____

8. Describe some periodic maintenance procedures that must be performed on the compressor and air source components of this part of the system.

Analysis:

1. What are some of the primary differences in a hydraulic source and a pneumatic system source?

2. What are some of the basic compressor types in operation today?

3. How does compressing air increase its pressure?

Activity 6-12—Pneumatic Pressure and Flow

Name ______________________________ Date ______________ Score ____________

Objectives:

When air passes through a pneumatic system, it normally encounters resistance that tends to reduce or lower the pressure between different locations. This lowering of pressure is primarily due to a form of friction that air encounters as it flows through the system. When airflow is reduced or stopped, it causes a change in friction. In a closed system without an airflow, the pressure is equalized throughout. When the flow of air increases, there is a corresponding increase in pressure drop. The length of the transmission flow path, the diameter of the tubing, number, and type of direction changes, and path restrictions all have some influence on the amount of pressure drop developed within a system.

In this activity, you will have an opportunity to construct a simple pneumatic system and alter the flow of air to see how it influences system pressure. You will be able to observe both an equalized system condition and the pressure drop of a heavy flow circuit. These conditions are very important considerations in the operation of a pneumatic system.

Equipment and Materials:

- Air supply source.
- Flow control valve.
- Pressure gauge—0–100 lb/in^2.
- Manual shutoff valve.
- Muffler.
- "T" connector.
- Assortment of quick-disconnect hoses.
- Safety glasses or goggles.

Procedure:

1. Construct the pneumatic circuit of **Figure 6-13**.

Figure 6-13. Pneumatic pressure and flow test circuit.

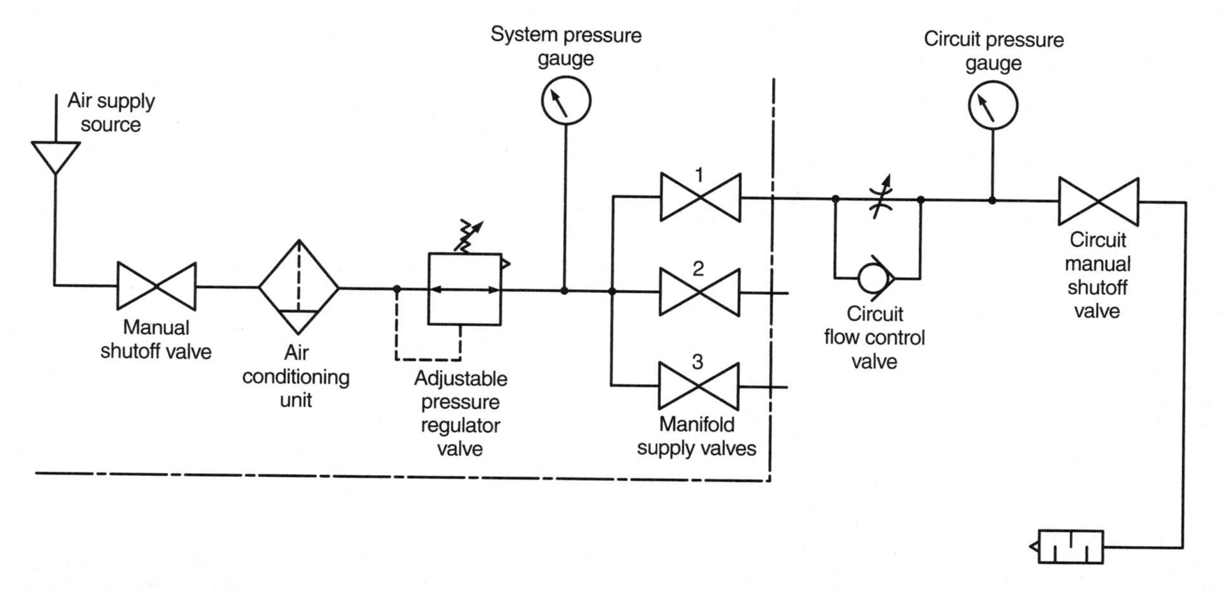

2. Close the manifold supply valves and turn on the manual shutoff valve. Adjust the pressure regulator valve to 15 lb/in^2 of system pressure. Close the circuit flow control valve and the circuit shutoff valve, then slowly open the manifold supply valve. The circuit pressure gauge should read zero at this time if everything is adjusted properly. If not, check the two circuit valves and make certain they are off.
3. Gradually open the flow control valve to produce a reading on the circuit pressure gauge. How does this reading compare with the system pressure?

   ____

4. Open the circuit manual shutoff valve. How do the two gauge pressures compare?

   ____

5. Turn off the circuit shutoff valve and add approximately six feet of hose between the muffler and the shutoff valve.
6. Turn on the manual shutoff valve again and observe circuit pressure. What influence does the additional hose have on the pressure reading?

   ____

7. Turn off the circuit shutoff valve and remove the muffler with the six-foot length of hose remaining.
8. Turn on the shutoff valve and observe the pressure reading. How does this compare with the value of Step 6?

   ____

9. Turn off the manifold supply valve and the manual shutoff valve of the air supply. Disconnect the circuit and return all components to the storage area.

Analysis:

1. What causes the pressure of a pneumatic circuit to drop at different locations?

   ____

   ____

2. What is meant by the term *system resistance*?

   ____

   ____

   ____

   ____

3. What are some of the things that alter the resistance to the flow of a pneumatic system?

   ____

   ____

   ____

   ____

Activity 6-13—Pneumatic Linear Actuators

Name ______________________ Date ______________ Score ____________

Objectives:

Pneumatic linear actuators are primarily used in industrial applications to lift, compress, hold, or position objects during different manufacturing processes. In order for this action to be achieved, air is fed into a chamber under pressure. A piston or ram in the chamber is then forced to move depending on the applied pressure. Pneumatic cylinders range in size from a fraction of an inch to several inches in diameter. Both single-acting and double-acting cylinders are used to achieve linear mechanical action in pneumatic circuits.

In this activity, you will have an opportunity to construct a simple linear actuator circuit to test the lifting capabilities of air. You will have an opportunity to observe this condition of operation with no load and under a loaded condition. Through this experience, you will become more familiar with basic pneumatic circuits and their operation. You will also have an opportunity to employ a flow control valve, which serves as a throttle that alters cylinder operating speed. This circuit modification is very important in many industrial applications today.

Equipment and Materials:

- Air supply source.
- Three-position, four-way control valve.
- Single-acting cylinder—1 1/8″ bore.
- Double-acting cylinder—1 1/8″ bore.
- "T" connectors (2).
- Pressure gauges—0–100 lb/in^2 (2).
- Flow control valve.
- Assortment of quick-disconnect hoses.
- Platform plate and 10-pound weight.
- Safety glasses or goggles.

Procedure:

Section A. Single-Acting Linear Actuators

1. Construct the linear actuator circuit of **Figure 6-14**. Note that the four-way valve has a plug attached to part of the valve that is not used.

Figure 6-14. Single-acting pneumatic linear actuator circuit.

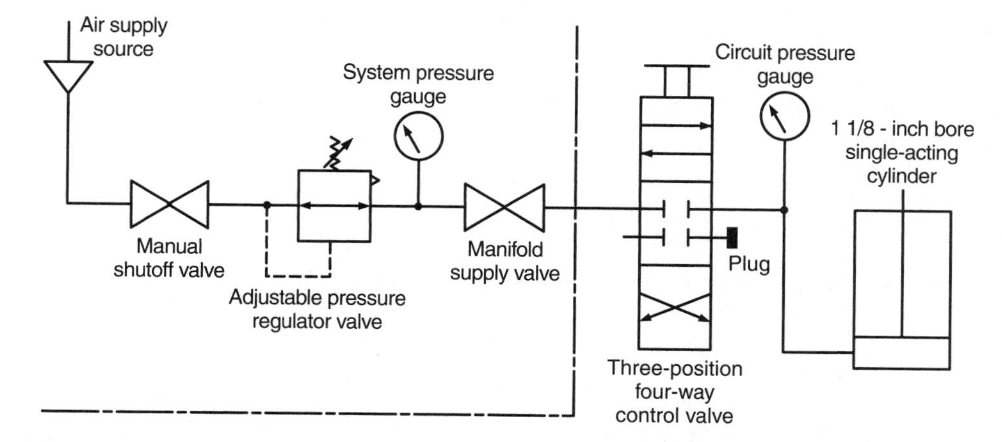

2. Turn off the manifold supply valve. Energize the air supply source and adjust the pressure regulator valve for 15 lb/in^2.
3. Turn on the manifold supply valve and switch the four-way valve to all three positions. Describe the action of the cylinder.

____

____

4. Place a weight on the cylinder or load it by holding it down with your hands. Run through the switching sequence again with an increased load. How does this influence the action of the cylinder?

____

5. Increase the system pressure to 40 lb/in^2 and repeat Steps 3 and 4. How does increased pressure alter the operation of the cylinder?

____

6. Turn off the manifold supply valve and insert the flow control valve as indicated in **Figure 6-15**.

Figure 6-15. Circuit from **Figure 6-14** with flow control valve added.

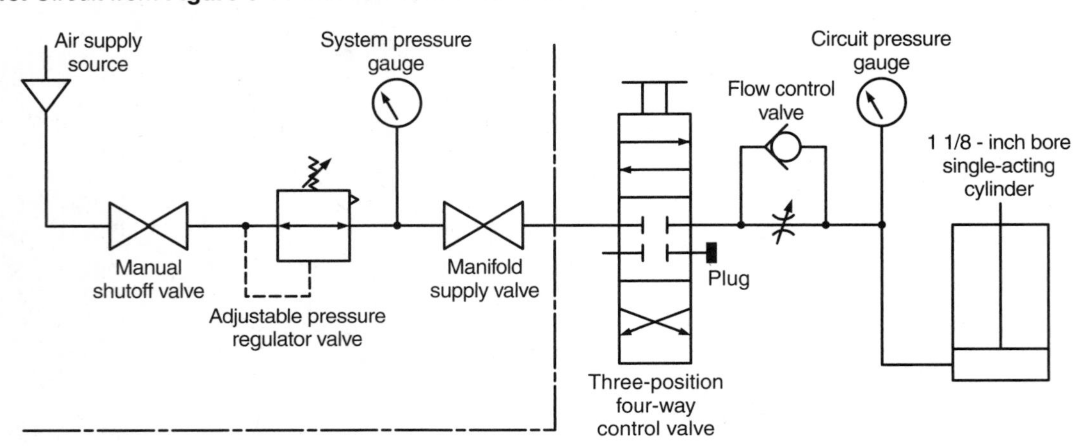

7. Turn on the manifold supply valve and switch the four-way valve to all three positions. In the retracting position, the flow control valve should be altered to regulate retracting speed. This type of valve serves as a throttle to regulate cylinder retraction.
8. Turn off the manifold supply valve and reverse the flow direction of the flow control valve.
9. Turn on the manifold supply valve and switch the four-way valve through its three positions. How does this alteration change the operation of the circuit?

____

____

10. Turn off the manifold supply valve and the air supply shutoff valve. Disconnect the circuit and return all parts to the storage area.

Section B. Double-Acting Linear Actuators

1. Construct the double-acting linear actuator circuit of **Figure 6-16**.
2. Turn off the manifold supply valve of the air supply and energize the source. Adjust the regulator pressure to 40 lb/in^2.

Name ______________________________

3. Turn on the manifold supply valve and switch the four-way valve to all three positions. Make a note showing the extended and retracted positions of the four-way valve. How do the pressure readings respond during operation of the cylinder?

____

____

4. Attach a platform to the top of the cylinder with the cylinder extending upward. Place a 10-pound weight on the platform and repeat Step 3. What influence does an increased load have on the operation of the circuit?

____

5. Turn off the manifold supply valve and place a flow control valve in the top cylinder line of the circuit. Connect as in **Figure 6-15** so the free flow is from the four-way valve to the cylinder with the throttled flow and from the cylinder to the four-way valve.

Figure 6-16. Double-acting pneumatic linear actuator circuit.

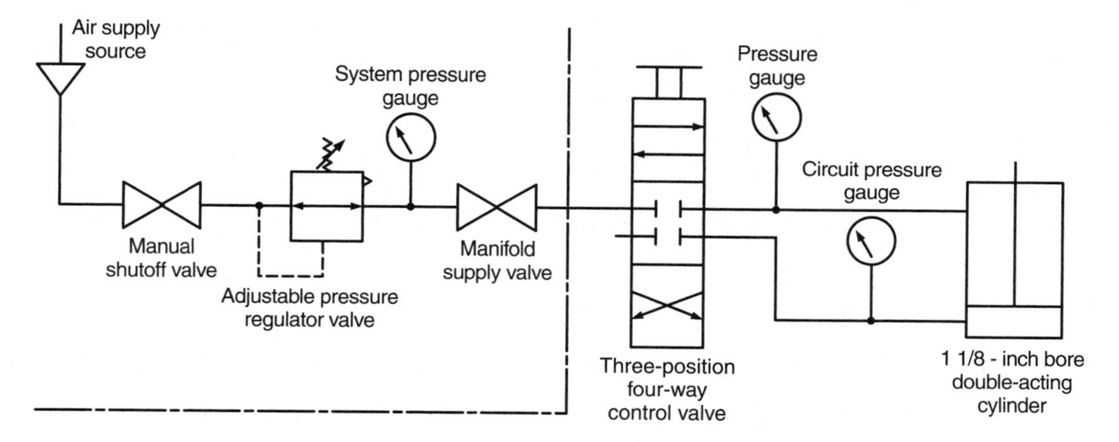

6. Turn on the manifold supply valve and switch the four-way valve to all three positions. In the retracting position, the flow control valve should be adjusted to produce a slow retracting action.
7. Turn off the manifold supply valve and switch the location of the flow control valve to the bottom line.
8. Turn on the manifold supply valve and switch the four-way valve through its three positions. How does the control valve alter the operation of the cylinder in this circuit location?

____

9. If two flow control valves are available, you can place one in each cylinder line and alter the operational speed of the valve in both the retracting and extending positions.
10. Turn off the manifold supply valve and the air supply shutoff valve. Disconnect the circuit and return all parts to the storage area.

Analysis:

1. What is an industrial application of a pneumatic circuit of the type studied in this activity?

____

____

2. What is a unique difference in this circuit operation and that of a similar hydraulic circuit?

____

____

3. Explain the operation of the throttling valve technique of cylinder speed control.

____

____

4. Make a circuit diagram for each of the three operating positions of the four-way valve with arrows showing the path of the airflow.

5. Why is the doubling-acting linear actuator considered to have better control than a single-acting cylinder?

____

____

6. Why does the four-way valve *not* require modification in this circuit as it did in **Figure 6-14**?

____

____

Activity 6-14—Pneumatic Rotary Actuators

Name ________________________________ Date ________________ Score ____________

Objectives:

Pneumatic rotary actuators are designed to produce rotary motion from air pressure. Air under pressure is applied to the rotating member, causing it to move. The air is then expelled from the outlet side. Reversing the direction of airflow causes the motor to rotate in the opposite direction. Pneumatic motors are lightweight, compact, and have a wide range of variable speed characteristics when compared with electric or hydraulic motors. These characteristics are well-suited for hand power tools, hoists, mixers, and in explosion-proof applications.

The horsepower and speed of a pneumatic motor can readily be changed by throttling the applied airflow. As the load applied to a pneumatic motor is increased, it normally causes a reduction in operational speed. This behavior is similar to that of an electric motor. With an increased load to an electric motor, however, the speed of rotation must quickly return to the normal running speed or it will be damaged by excessive heat. Excessive loading of a pneumatic motor will not adversely damage the motor because the airflow has a significant cooling effect. It can be run at alternate speeds without adverse damage to the motor for prolonged periods of time.

Pneumatic motors are energized by air, which means they can be used in potentially explosive environments. This characteristic is extremely important in many industrial applications today.

The motor used in this activity produces up to 1/3 horsepower and operates from 500 to 10,000 rpm. It is primarily designed to respond to pressures below 100 lb/in^2. The amount of airflow needed to operate this motor is quite high, however. As a rule, the overall operating efficiency of a pneumatic motor is less than 20% for continuous operation. Air motors are rarely used when operational efficiency is important. They are used where the variable speed characteristic or explosion-proof advantage outweighs efficiency.

In this activity, you will construct a simple pneumatic motor circuit and test the loading effect on applied pressure and speed. You will then test the direction of rotation and see if loading and speed are altered. Through this activity, you will become more familiar with the operation of a pneumatic motor and see how rotary actuators can be harnessed to do work.

Equipment and Materials:

- Air supply source.
- Flow control valve.
- Three-position, four-way control valve.
- Reversible pneumatic motor—Gast No. 1AM vane type or equivalent.
- Muffler.
- Assortment of quick-disconnect hoses.
- Pressure gauges—0–100 lb/in^2 (2).
- Safety glasses or goggles.

Procedure:

Section A. Rotary Actuator Characteristics

1. Construct the rotary actuator circuit of **Figure 6-17**.
2. Turn off the manifold supply valve, then energize the air supply. Adjust the pressure regulator valve for 30 lb/in^2 of system pressure.
3. Close the flow control valve, then open the manifold supply valve. Slowly open the flow control valve while observing the circuit pressure gauge. At what pressure does rotation occur?

   Rotation occurs at ________________ lb/in^2.
4. Note the direction of motor rotation.

   Direction of rotation is ________________ (clockwise or counterclockwise).

Figure 6-17. Pneumatic rotary actuator circuit.

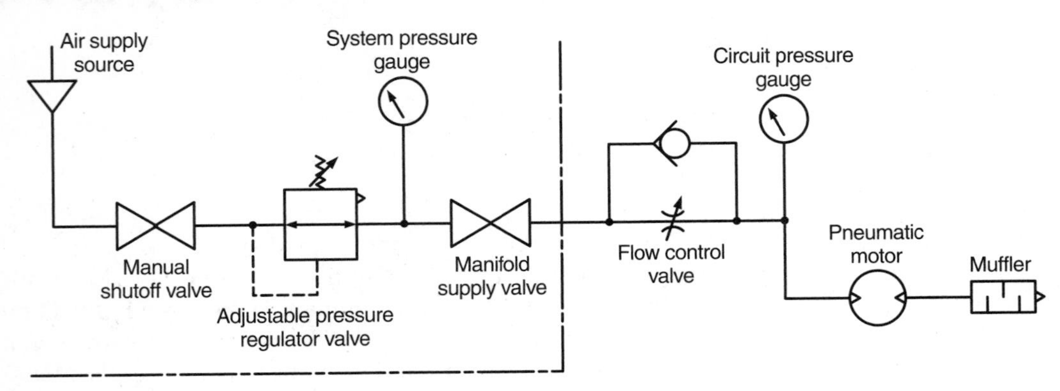

5. Gradually increase the flow while observing the circuit pressure. What is the maximum pressure observed?

   Maximum pressure = ________________ lb/in^2.

   Note that when the pressure increases, the speed increases accordingly.

6. Connect a tachometer to the motor shaft and test the no-load speed.

   Speed = ________________ rpm.

7. Load the motor by carefully wedging a piece of wood between the rotating shaft and the base of the motor. How does loading influence speed and circuit pressure?

   ____

8. Momentarily turn off the manifold supply valve and wait for the motor to stop rotation.

9. Adjust the pressure regulator valve to 70 lb/in^2 of system pressure. Close the flow control valve.

10. Turn on the manifold supply valve and gradually turn on the flow control valve. The motor should be loaded with the wooden wedge again. Increase the flow until rotation occurs. What circuit pressure is needed to produce initial rotation of the motor under a loaded condition?

    Pressure = ________________ lb/in^2.

    How does this compare with the unloaded condition of Step 3?

    ____

11. Turn off the manifold supply valve and turn off the air supply source. Disconnect the circuit and return all components to the storage area.

Section B. Rotary Actuator Directional Control

1. Construct the rotary actuator circuit of **Figure 6-18**.

Figure 6-18. Pneumatic rotary actuator directional control circuit.

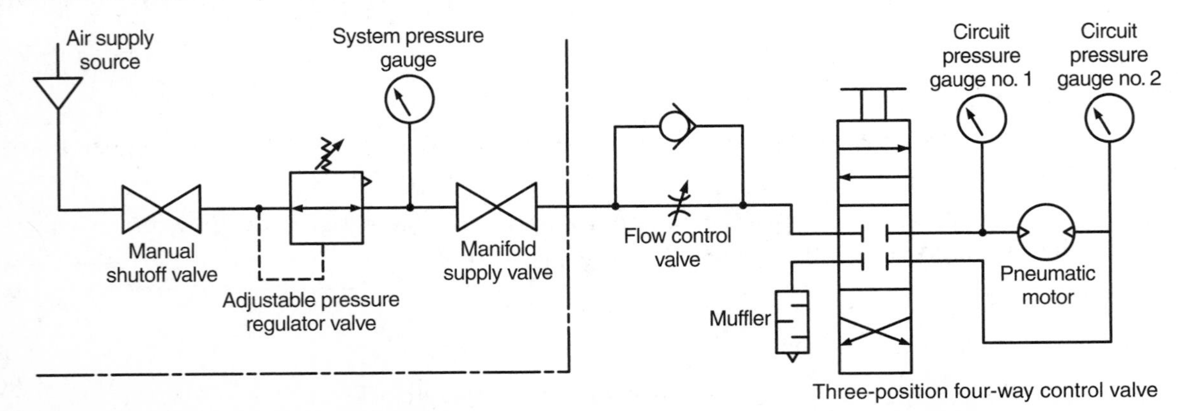

Name ___________________________________

2. Turn off the manifold supply valve and energize the air supply source. Adjust the pressure regulator valve to 70 lb/in^2 of system pressure.
3. Turn off the flow control valve, then turn on the manifold supply valve. Switch the four-way directional control valve to each of the three positions. Then switch the four-way valve to one of its extreme positions.
4. Carefully adjust the flow control valve to produce a slow rotation of the motor. Make a note indicating the direction of rotation produced by this switch position.
5. Record the pressure indicated by gauges 1 and 2.

   Gauge No. 1 = ____________________ lb/in^2.

   Gauge No. 2 = ____________________ lb/in^2.
6. Switch the four-way control valve to the center or "off" position. Wait momentarily until rotation stops. Then switch the four-way control valve to the other extreme position. Make a note indicating the direction of rotation produced by this switch position.
7. Record the pressure indicated by gauges 1 and 2.

   Gauge No. 1 = ____________________ lb/in^2.

   Gauge No. 2 = ____________________ lb/in^2.
8. Increase the speed of rotation by opening the flow control valve. With a tachometer, measure the rotational speed. (Note: Some tachometers will only indicate speed in one rotational direction. Test the tachometer before making this measurement.)

   Speed = ____________________ rpm.
9. Switch the four-way control valve to the "off" position until the motor stops. Then switch it to produce rotation in the opposite direction. Observe the note in Step 8.

   Speed = ____________________ rpm.

   Does the motor run equally well in either direction?

   ____
10. If time permits, repeat Steps 1–9 after wedging a piece of wood between the motor shaft and its base as a loading device.
11. Turn off the manifold supply valve and the supply shutoff valve. Disconnect the circuit and return all parts to the storage area.

Analysis:

1. What influence does loading have on motor starting torque and the applied starting pressure?

   ____

   ____
2. How does loading alter the rotational speed of a pneumatic motor?

   ____

   ____
3. Of what significant advantage would a pneumatic motor have in industry?

   ____

   ____

4. How would the pressure gauges of this circuit respond if the muffler were removed?

____

5. What are some of the common types of pneumatic motors in operation today?

____

____

6. What kind of pneumatic motor is used in this activity?

____

7. How does a pneumatic motor compare with an electric motor?

____

____

Activity 6-15—Reed Switches

Name ______________________ Date ______________ Score __________

Objectives:

Reed switches are devices that respond to a controlled magnetic field. They are designed to close or open when exposed to either a permanent magnetic field or to an electromagnetic field. The reed switch is representative of various types of switching devices that may be utilized with robotic systems.

The contacts of a reed switch are housed inside a hermetically sealed glass tube. When actuated, contact sparks are isolated from the outside environment. Industrial applications of this device are quite numerous in explosive areas and dirt-prone environments. They are used to verify Z axis movement on some robots and other applications involving robotic systems.

A reed switch contains two flat metal strips, or reeds, housed in a hollow glass tube filled with an inert gas. When the reeds are exposed to a magnetic field, they are forced together, making or breaking contact depending on their design. The normally closed switch breaks contact when exposed to a magnetic field. Normally open contacts, by comparison, are forced closed when exposed to a magnetic field.

In this activity, you will construct a simple electrical circuit and control its action by changes in magnetic field strength. Through this experience you will gain some insight into the operation, sensitivity, and control action of a reed switch.

Equipment and Materials:

- Dc power supply—0V–5V, 1.0 A.
- Ac power source—6.3V, 60 Hz.
- Reed switch—GE-X7 or equivalent.
- Bobbin-wound, reed switch coil.
- Reed switch magnet.
- No. 47 lamp.
- SPST Switch.
- Circuit construction board.

Procedure:

1. Construct the reed-switch circuit of **Figure 6-19**.

Figure 6-19. Reed switch test circuit for observing permanent magnet actuation.

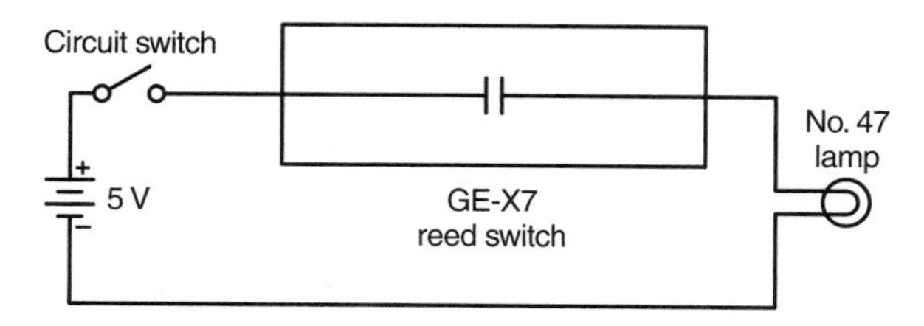

2. Turn on the circuit switch and move the small permanent magnet near the reed switch. Try the magnet at different orientations with the reeds. Which type of magnet orientation produces the best control capability?

____

3. Place the reed switch in the center of the bobbin-wound coil. Connect the circuit of **Figure 6-20** to test the electromagnetic action of the reed switch.

Figure 6-20. Reed switch test circuit for observing electromagnetic actuation.

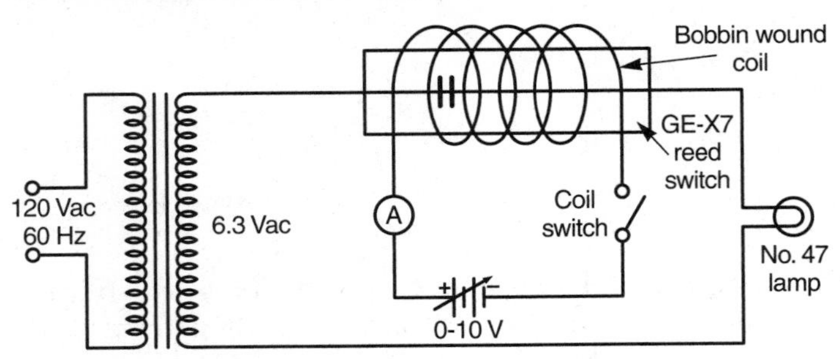

4. Turn on the ac power source and the dc coil-control circuit.
5. Starting at zero volts dc, gradually increase the dc source voltage until the reed switch is actuated. The meter should be set at a high range until an approximate actuating value is determined. Run at least two trial tests to determine the actuating current needed to energize the switch.

   Dc actuating current = ____________________ milliamperes.
6. In some applications, the actuating coil is used to increase the sensitivity of the reed switch by producing a partial field. Increase the dc coil current close to the actuating value. Then place the permanent magnet near the coil to actuate the reed. In this case, the switch can have some degree of variable sensitivity.
7. Turn off the ac and dc power sources for the circuit of **Figure 6-20**. Use 5V dc to supply the coil. The primary side of the coil should be connected to a variable ac transformer.
8. Turn on both the ac and dc sources, then increase the ac voltage applied to the coil. Measure the applied ac voltage. Do not increase the voltage to a value that will cause the reed coil to overheat. How does the reed switch respond to ac compared with dc?

   ____
9. Turn off the ac and dc power sources and disconnect the circuit. Return all components to the storage area.

Analysis:

1. How could a reed switch be used to control a pump motor in a liquid sump tank?

   ____

   ____
2. Make a sketch of this circuit.
3. How could a reed switch be used to control a circuit in an explosive area without danger of an explosion?

   ____

   ____

Activity 6-16—Liquid Level Control

Name ______________________________ Date ______________ Score __________

Objectives:

Many industrial processes rely on liquid level measurement. Sometimes variables such as fuel supply are monitored continuously. There are several techniques that may be used to measure liquid level. Liquid level control is a type that might be used in conjunction with robotic systems.

The measurement of liquid level is easy to accomplish by using transducers. Level changes result in the displacements of the top surface of the liquid. Many types of transducers may be used to measure liquid level. Resistive transducers may be used to measure the level of a conductive solution. A capacitive transducer may be used along with a movable plate whose position is determined by the level of the liquid. Photoelectric methods, radioactive methods, and ultrasonic methods may also be employed.

A simple type of level controller is the ball float system. This system uses a ball float to operate a lever. The lever is connected to a valve, which regulates liquid flow rate. Chemical industries commonly use differential pressure controllers to control the level of volatile liquids. The liquid pressure is proportional to its level in the enclosed container.

Level control may be accomplished by placing a light source at the same height above a conveyor line as the desired fill level of a container. In the illustration of **Figure 6-21**, containers on a conveyor line are positioned under a liquid dispenser. When the container is in position, the actuator causes the dispenser to allow liquid to pass into the container. When the liquid reaches the level of the light source, the light beam is interrupted. With no light striking its surface, the detector will cause a relay to activate. The activated relay will, in turn, cause the actuator on the dispenser to close. When another container is in position, the actuator will open once more. This liquid level control system ensures a uniform level of liquid in each container. In this activity, you will construct and test a photoelectric circuit, which could be used to measure liquid level.

Figure 6-21. Method of liquid level control.

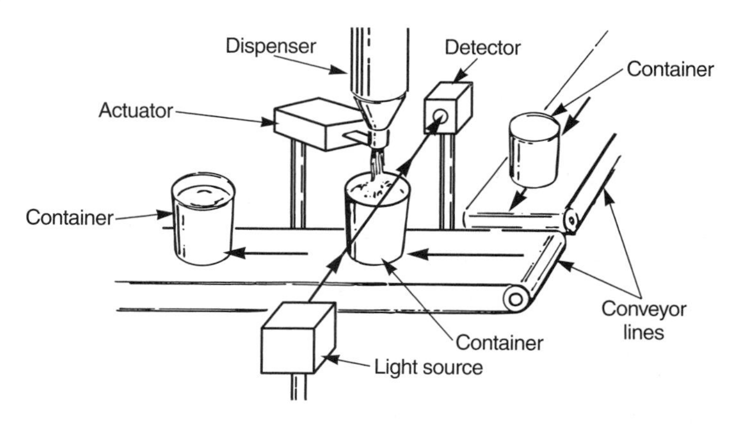

Equipment and Materials:

- Variable dc power supply.
- Light-dependent resistor (VAC 54 or equivalent).
- Electronic multifunction meter.
- Container (glass or plastic).
- 12V, 1250 Ω relay (Guardian 1335-2C-120D or equivalent).
- 60 W lamp with holder.
- 7 W lamp with holder.
- SPST switch.
- 120V ac power source.

Procedure:

1. Construct the photoelectric liquid level control of **Figure 6-22**.
2. Plug the light source into a 120V ac power outlet and close the circuit switch. Position the light source near the bottom of the glass container before it is filled with water. Slide the paper tube around the light-dependent resistor (ldr) and place the open end against the container.

Figure 6-22. Photoelectric liquid level control.

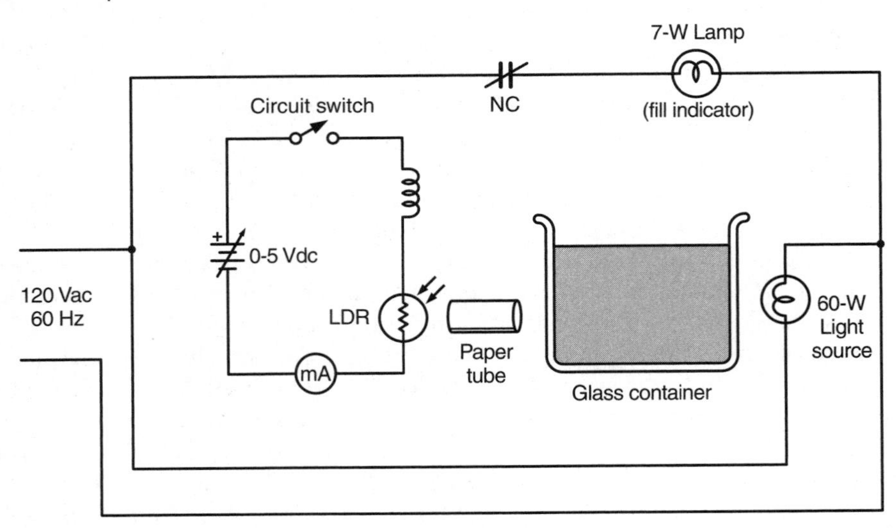

3. If the circuit is operating properly, the fill indicator lamp will light when the relay is actuated.
4. Adjust the dc source to alter the sensitivity of the circuit. Do not exceed 15V dc. The circuit should be able to detect a pencil passing in front of the paper tube.
5. Carefully fill the container with water until the indicator turns off. Avoid pouring water directly in front of the tube window area. You may need to try several trial runs to get the sensitivity to a level where the circuit will respond properly.
6. After the sensitivity has been adjusted, drain or siphon water from the container until the fill indicator is actuated again.
7. Test the liquid level control circuit two or three times to ensure it operates properly.
8. This concludes the activity. Disassemble and return all components to the storage area.

Analysis:

1. Discuss the operation of the circuit used in this activity.

____

____

____

____

2. What are some other types of circuits that could be used for liquid level control?

____

____

____

____

Activity 7-1—Proximity Detectors

Name ______________________ Date ______________ Score ____________

Objectives:

Proximity detectors are used to trigger an alarm or turn on a circuit by detecting physical change within a designated area. These circuits are designed to respond in some way to a change in capacitance. This change usually alters the frequency of an oscillator so the output can be amplified and used to trigger a load control device. Typically the circuit contains an oscillator, a tuned circuit, an amplifier, and a trigger device. This type of sensor is representative of the type of sensors that might be used in robotic system applications.

One type of proximity detector utilizes the loaded oscillator principle. In this circuit, the LC components (inductor and capacitor) of the oscillator are shunted to ground by an external capacitance. This capacitance is usually called the *sensor plate*. Moving a hand or producing a physical change near the sensing plate upsets the electrostatic field of the capacitance. This, in turn, alters the frequency of the oscillator or may load down the circuit in such a way as to quench the oscillation process. Liquid level detection, automatic thickness gauge testers, intruder alarm systems, counters, and switching circuits employ this type of detector.

In this activity, you will be able to construct a loaded oscillator proximity detector. The oscillator is a Colpitts type with split capacitors in the transistor collector circuit. The output of the oscillator is rectified and applied to a voltage comparator circuit where it is amplified by an op amp. The amplified output is used to trigger a load control device. The 5 kΩ resistor in the comparator circuit is used to establish a balanced condition for detection. When the oscillator senses a change, it loads down and stops oscillating. This creates an imbalance, which is then detected by the comparator circuit.

Through this activity, you will be able to test the operation of a loaded oscillator circuit and observe typical waveforms. The test procedure used to analyze circuit operation is also used to troubleshoot the circuit. Through this testing experience you will become familiar with the oscillator loading principle and be able to follow a logical troubleshooting procedure.

Equipment and Materials:

- 0–10V, 1 A dc power supply.
- LM3900 op amp.
- 2N3397 transistor.
- 10 mH coil.
- 1N4004 diode.
- Light-emitting diode.
- 0.1 μF, 200V dc capacitor.
- 0.01 μF, 100V dc capacitor.
- 100 pF, 100V dc capacitor.
- 47 pF, 100V dc capacitor.
- 1 MΩ, 1/4 W resistors (2).
- 68 kΩ, 1/4 W resistor.
- 10 kΩ, 1/4 W resistors (2).
- 6.8 kΩ, 1/4 W resistor.
- 5 kΩ, 2 W potentiometer.
- 1 kΩ, 1/4 W resistor.
- 470 Ω, 1/8 W resistor.
- 500 kΩ, 2 W potentiometer.
- SPST toggle switches (2).
- 6″ × 6″ sensor plate (metal).

Optional according to the selected load control circuit.

- 2N3397 transistor.
- 12V, 125 Ω relay (Guardian 1335-2C-120D or equivalent).
- No. 47 lamp with socket.
- 47 Ω, 1/2 W resistor.
- SCR (MCR 12-D or equivalent).
- 220 Ω, 1/2 W resistor.
- 6.3V, 60 Hz ac source.

Procedure:

1. Construct the proximity detector of **Figure 7-1**.

Figure 7-1. Proximity detector.

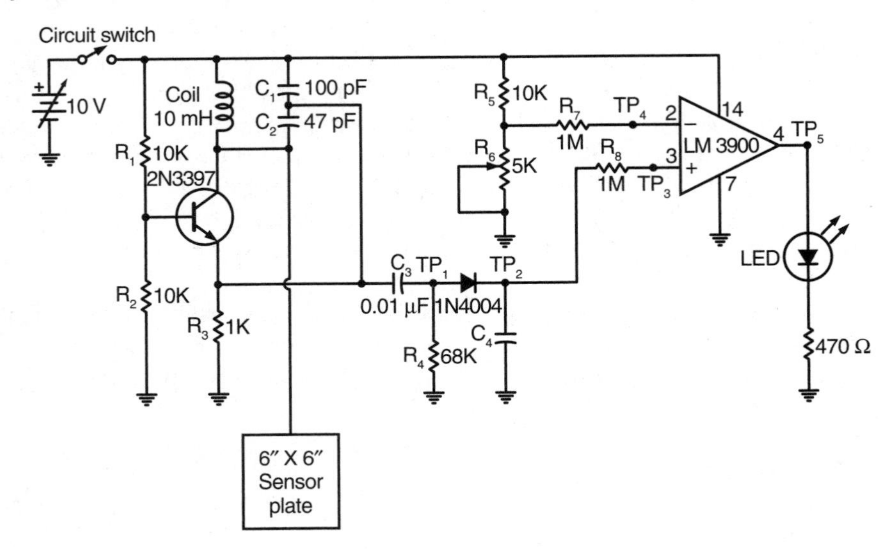

2. Close the circuit switch and adjust the 5 kΩ potentiometer for illumination of the LED. There should be a specific point where the LED turns on. Increasing resistance will cause it to remain on, while decreasing the resistance causes it to turn off. The point just before turn-on represents the most sensitive detection setting of the potentiometer. Adjust the potentiometer several times to locate the most sensitive point.
3. If the circuit is not operating properly, perform Steps 4, 8, 9, 10, and 11 first to determine which part of the circuit is malfunctioning. If the circuit is operating satisfactorily, follow the procedure steps in order and record the indicated data.
4. Prepare an oscilloscope for operation and connect it to test point 1. Make a sketch of the observed waveform, **Figure 7-2**.

Figure 7-2. Observed waveform at test point.

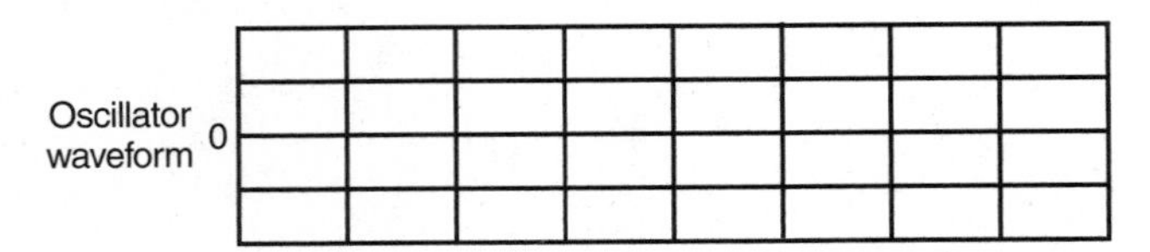

5. While observing the oscilloscope, touch the sensor plate. What influence does this have on the waveform?

Name ____

6. Move your hand over the sensor plate while observing the oscilloscope. What influence does this have on the waveform?

   ____

   ____

7. Determine the frequency of the oscillator with the oscilloscope or a frequency meter.

   Oscillator frequency = ________________ Hz.

8. Measure and record the voltages at the emitter-base and collector of the oscillator transistor Q_1.

   Emitter voltage = ________________ V.

   Base voltage = ________________ V.

   Collector voltage = ________________ V.

   Compare your measurements with those indicated on the schematic diagram.

9. Measure and record the dc output voltage at test point 2.

   Unloaded oscillator output voltage = ________________ V.

10. Touch the sensor plate and record the measure voltage at test point 2. This represents the loaded output voltage of the oscillator.

    Loaded output voltage = ________________ V.

    The output of the oscillator will change approximately 2V between the loaded and unloaded conditions of operation.

11. Measure and record input voltages at TP_3 and TP_4 when the LED is on. This indicates a balanced condition.

    TP_3 = ________________ V.

    TP_4 = ________________ V.

12. Measure and record the voltages at TP_3 and TP_4 when the circuit is unbalanced.

    TP_3 = ________________ V.

    TP_4 = ________________ V.

13. Measure and record the output voltage of the circuit at test point TP_5 for the balanced and unbalanced conditions. The total change in voltage is ________________ V.

14. Select one of the alternate load control circuits in **Figure 7-3** and construct it. Connect it to the output of the proximity circuit.

Figure 7-3. Alternate load control circuit.

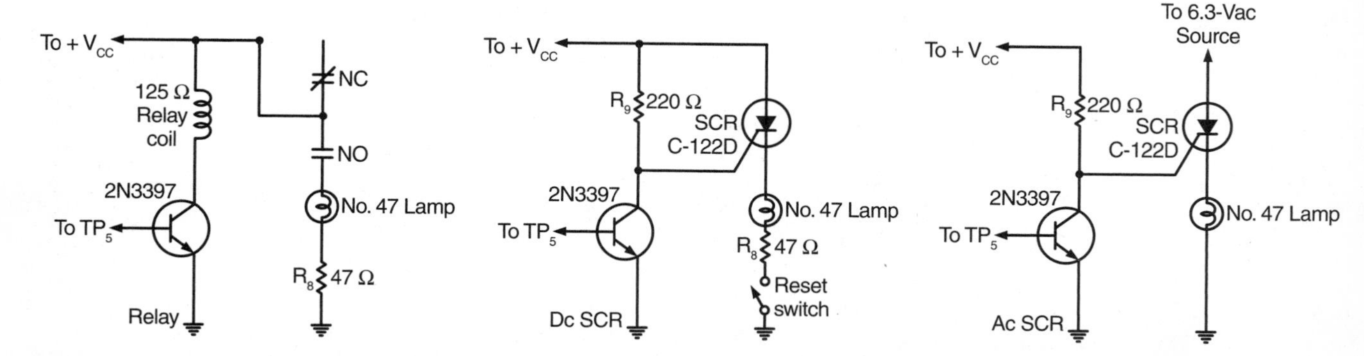

15. Close the circuit switch, adjust the balance potentiometer, and test the control capability of the completed load control circuit.
16. Open the circuit switch. Disconnect the circuit and return all parts to the storage area.

Analysis:

1. Describe what is meant by loaded and unloaded oscillator conditions.

____

____

____

2. What is the function of the IC op amp in the proximity circuit of **Figure 7-1**?

____

____

3. Draw a block diagram representing the major parts of the proximity circuit constructed in Step 15.

Activity 7-2—LVDT Detector

Name ______________________ Date ______________ Score ____________

Objectives:

Frequently, it is not enough to measure only the amount of the displacement of an object. Often the direction of displacement, as well as the amount of displacement, must be indicated. When this becomes the case, a detector circuit similar to that shown in **Figure 7-4** can be used in conjunction with the output of the linear variable displacement transformer (LVDT).

Figure 7-4. LVDT detector circuit.

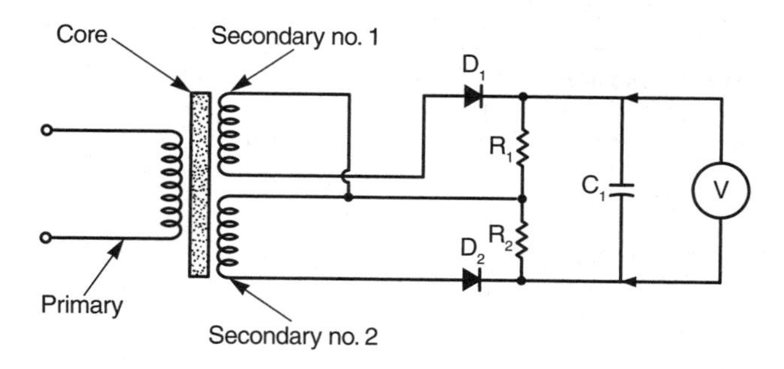

In this sensor circuit, the diodes conduct equally when the movable core is in the null position. This results in an equal and opposite voltage drop across R_1 and R_2, causing the readout device (voltmeter) to indicate zero displacement. When displacement occurs, one diode will conduct more than the other, causing an upset in the balanced voltage across R_1 and R_2. The result is an output voltage, as measured by the voltmeter, whose polarity indicates the direction of displacement and whose magnitude indicates the amount of displacement. In this activity, you will examine a detector circuit that might be used with robotic systems or other applications utilizing sensors.

Equipment and Materials:

- Digital multimeter.
- Air-core coils, 100 turns, A.S. No. 16 wire (2).
- Air-core coil, 200 turns, A.S. No. 24 wire.
- 6″ (15 cm) steel core, 3/4″ (1.91 cm) diameter.
- 6V ac power supply.
- 1N4004 diodes (2).
- Resistors—1 kΩ (2).
- Capacitor—1 μF, 25V dc.
- Connecting wires.
- SPST switch.

Procedure:

1. Construct the circuit illustrated in **Figure 7-5**.
2. Close the SPST switch and slide the movable core until the voltmeter indicates zero.
3. Moving the core in the direction and by the amount indicated in **Figure 7-6**, complete the table by recording the voltage output of the detector circuit along with the proper voltage polarity.

Figure 7-5. LVDT detector test circuit.

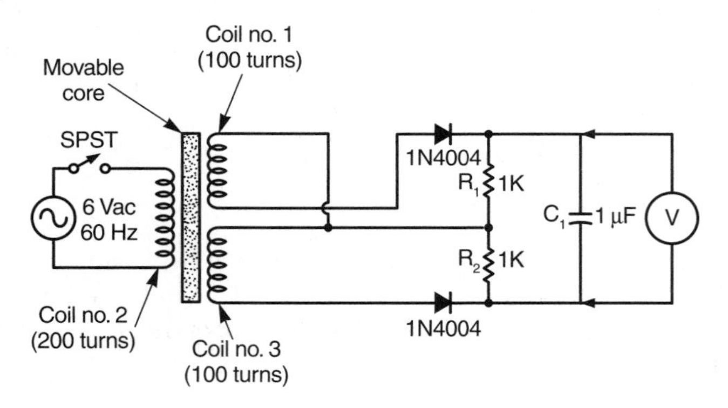

Figure 7-6. Output polarity and magnitude of LVDT test circuit.

Direction of movement from null position	Amount of movement Inches	Centimeters	Voltage output	Polarity (+ or –)
Left	1/16	0.159		
Left	1/8	0.318		
Left	3/16	0.477		
Left	1/4	0.635		
Left	5/16	0.793		
Left	3/8	0.953		
Return to Null Position				
Right	1/16	0.159		
Right	1/8	0.318		
Right	3/16	0.477		
Right	1/4	0.635		
Right	5/16	0.793		
Right	3/8	0.953		

4. How do the voltages generated by moving the core to the left compare with the voltages generated when the core is moved to the right?

____

____

5. Ideally, what is the minimum displacement of the movable core that would result in a voltage output?

____

____

Name ______________________________

Analysis:

1. What is meant by coils of an LVDT being connected in series to oppose?

 ____

 ____

2. Explain how the detector circuit used in conjunction with the LVDT enables the direction of displacement to be measured.

 ____

 ____

 ____

3. When might it be necessary to measure direction, as well as amount of displacement?

 ____

 ____

4. How would the null position of the core of the LVDT be affected if R_1 of the detector circuit were a value different from R_2?

 ____

 ____

Activity 7-3—Photoelectric Sensors

Name ______________________ Date ______________ Score ____________

Objectives:

Photoconductive sensors are designed to produce changes in their electrical conductivity when variations of light energy occur. These devices are also called photoresistive, since their resistance varies in inverse proportion to their conductivity. The cadmium sulfide (CaS) cell shown in **Figure 7-7** is a common type of photoconductive cell. When exposed to varying intensities of visible light, the cadmium sulfide cell will change resistance. An increase in light energy falling onto its surface will increase the conductivity of the cell. The cell is highly sensitive to variations of light intensity. It is typically used in alarm and relay control systems as a type of sensor.

Figure 7-7. Cadmium sulfide photoconductive cell.

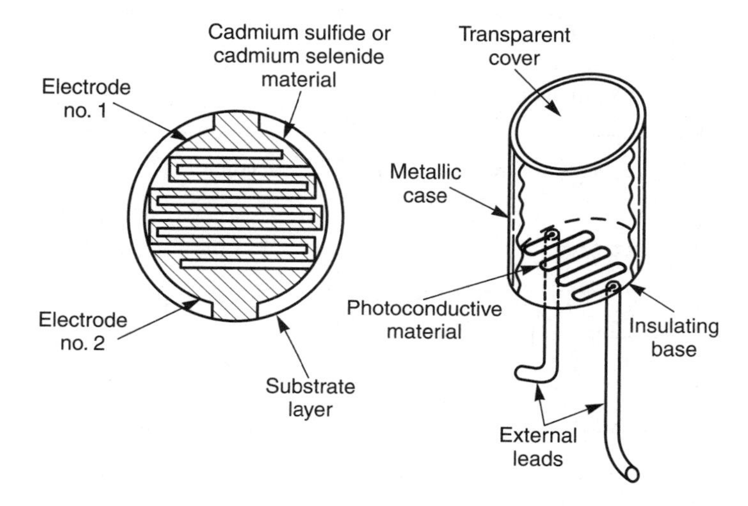

When light strikes the surface of a photoconductive cell, valance electrons of the semiconductor material are released from their atomic bonds. When electrons are released, the resistance of the material decreases. We may also say that the material becomes more conductive. If the light intensity is increased, more electrons will be released and the material becomes more conductive. Resistance of photoconductive devices may range from several megohms in darkness to 50–100 Ω in fairly intense light. In this activity, you will observe the characteristics of a photoconductive cell. The photoconductive cell is a popular type of photoelectric transducer that is used for various sensing applications.

Equipment and Materials:

- Photoconductive cell—VAC 54.
- Variable ac power supply.
- Multifunction meter.
- Lamp—60 W.
- Connecting wires.

Procedure:

1. Assemble the circuit shown in **Figure 7-8**. Connect the meter across the photoconductive cell so it will measure resistance changes with changes in light intensity. The 60 W lamp is connected to a 0–120V ac power source.
2. Record the resistance of the photoconductive cell with room light.

   R = ______________ Ω

Figure 7-8. Photoconductive cell test circuit.

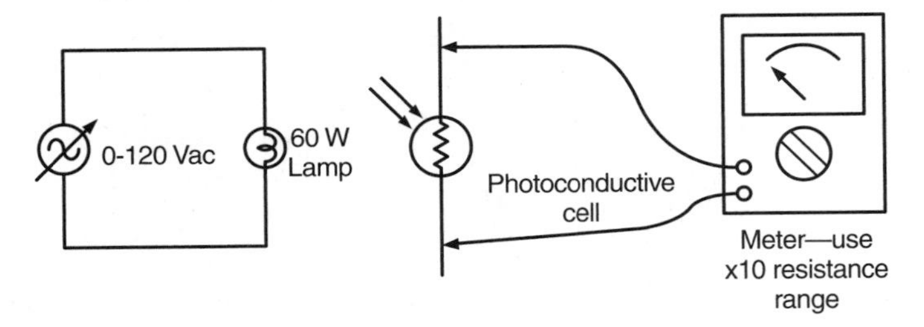

3. Cover the top of the cell with your finger and record the resistance without light.

   R = ________________ Ω

4. Place the 60 W lamp within 1/2″ (1.27 cm) of the top of the cell. Adjust the variable ac source to zero volts.
5. Turn on the ac source and complete the table in **Figure 7-9** by adjusting the source voltage as indicated.
6. This concludes the activity. Disassemble and return all components to the storage area.

Figure 7-9. Photoconductive cell characteristics.

Light source voltage	Resistance (Ω)
0	
10	
20	
30	
40	
50	
60	
70	
80	
90	
100	
110	
120	

Analysis:

1. What is the range of resistance change of the photoconductive cell used in this activity?

   Range = ________________ to ________________ Ω

2. Discuss the characteristics of a photoconductive cell.

____

____

____

____

Activity 7-4—Temperature Sensors

Name ______________________ Date ______________ Score ____________

Objectives:

Temperature sensing circuits are widely used in industry as system control elements. Heat sensors, which are by far the most popular of all sensors, are commonly found in alarm circuits that detect changes in temperature at remote locations. Machinery, electronics equipment, and measuring instruments also employ temperature sensors to detect unusual operating conditions. The thermistor is a very popular sensor element for temperature detection that might be used with robotic applications.

Thermistors are primarily classified as temperature sensitive resistors with an operating range of from –382°F (–230°C) to 1202°F (650°C). When used in bridge circuits with high-gain amplification, thermistors can detect temperature changes as small as 0.001°.

The physical makeup of a thermistor includes a mixture of nickel, manganese, and cobalt oxides formed into a piece of semiconductor material. These oxides are mixed together and fired to form a coherent nonporous material. The mixture is then formed into a variety of shapes. In this activity, a tiny piece of the ceramic material is formed into a bead and enclosed in glass. Changes in temperature cause the resistance of the thermistor to produce a wide range of values. Typically, thermistors have a negative temperature coefficient. This means that an increase in temperature causes a decrease in resistance. Metal, by comparison, has a positive temperature coefficient. This means that a rise in temperature causes a corresponding increase in resistance.

In this activity, you will build a temperature detecting bridge circuit. The output of the bridge is then fed into an op amp for high-gain amplification and increased sensitivity. Any imbalance in the bridge is detected by the amplifier and is used to trigger an output circuit. Circuits of this type are typically found in alarm circuits that function as heat detectors and might be used with robotic and assembly applications in industry.

Equipment and Materials:

- Split dc power supply: ±9V at 1 A, or two 9V batteries.
- μA741C op amp.
- Thermistor (Fenwal gB 32J2 or equivalent).
- 200 kΩ, 1/4 W resistor.
- 10 kΩ, 1/4 W resistors (2).
- 5 kΩ, 2 W potentiometer.
- 2.7 kΩ, 1/4 W resistor.
- 1 kΩ, 1/4 W resistor.
- 47 Ω, 1/4 W resistor.
- No. 47 lamp with socket.
- IC circuit construction board.
- SCR (GE C-122D or equivalent).
- Electronic multifunction meter.
- Decade resistance box (optional).
- SPST toggle switch.

Procedure:

1. Construct the thermistor bridge circuit of **Figure 7-10**. Note that a voltmeter is used to detect the amplified output of the circuit.
2. Before connecting the thermistor into the circuit, measure its resistance. Try to avoid touching the glass part of the thermistor when attaching the test leads. With the ohmmeter connected, the resistance should stabilize after a few seconds. The stabilized resistance is ______________ Ω.

Figure 7-10. Thermistor bridge circuit.

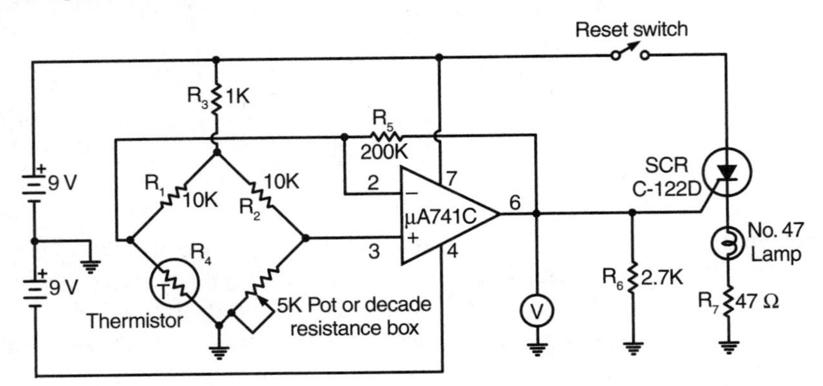

3. While observing the ohmmeter, grasp the glass bead between your index finger and thumb. How does body temperature influence the resistance of the thermistor?

___

___

4. Connect the thermistor to the bridge circuit as indicated. The reset switch should be in the off position and the electronic multifunction meter should be adjusted to the zero center position. It should be in the 50V or equivalent range.
5. Turn on the split power supplies and energize the circuit. Balance the bridge by adjusting the potentiometer to produce a zero indication on the meter. A decade resistance box can be used in place of the potentiometer.
6. After the bridge has been nearly balanced, you may switch the voltmeter to a lower range to improve the balancing accuracy.
7. To test the sensitivity of the circuit, place your finger near the glass bead of the thermistor while observing the voltmeter. If the circuit is working properly, the meter should deflect upscale or in the positive direction. Within a few seconds it should return to the balanced indication when the thermistor has reached its stabilized resistance. If you touch the glass bead with your finger, it usually takes longer for it to return to the balanced state.
8. You may want to try blowing on the thermistor or placing it near some heat-producing source. Avoid temperatures over 302°F (150°C).
9. Close the reset switch of the SCR. Place your finger near the thermistor. While observing the voltmeter, see how much voltage is needed to trigger the SCR into conduction. You must wait a few seconds for the thermistor to stabilize before resetting the SCR.
10. Try several sources of heat to turn on the sensor circuit.
11. Turn off the power supply and disconnect the circuit. Return all components to the storage area.

Analysis:

1. What type of temperature coefficient does the thermistor used in this activity have?

___

___

2. Explain how a change in thermistor resistance causes the bridge circuit to be imbalanced.

___

___

3. Why is it advantageous to use an op amp with the bridge circuit of this activity?

___

___

Activity 7-5—Thermocouple Applications

Name ______________________ Date ______________ Score ____________

Objectives:

Thermocouples are frequently used to measure temperatures in an industrial setting. Due to the relatively low voltage output associated with most thermocouples, amplification circuits are used to increase this output as well as to increase sensitivity. The resulting output of the amplification circuit is used to drive or activate a readout device.

In this activity, you will see the thermocouple used to control the conductivity of an FET. The conductivity of the FET, in turn, controls the action of a single stage transistor amplifier and, therefore, it controls the current flow through a multimeter that is used as the readout device.

Equipment and Materials:

- Type J thermocouple.
- Digital multimeter.
- Resistors—560 Ω, 5.6 kΩ.
- 0–5V dc power supply.
- 10V dc power supply.
- GE-FET-1 field-effect transistor.
- 2N2405 npn transistor.
- Connecting wires.
- 660 W heat cone.
- 120V ac power supply.

Procedure:

1. Construct the circuit shown in **Figure 7-11**.

Figure 7-11. Type J thermocouple FET circuit.

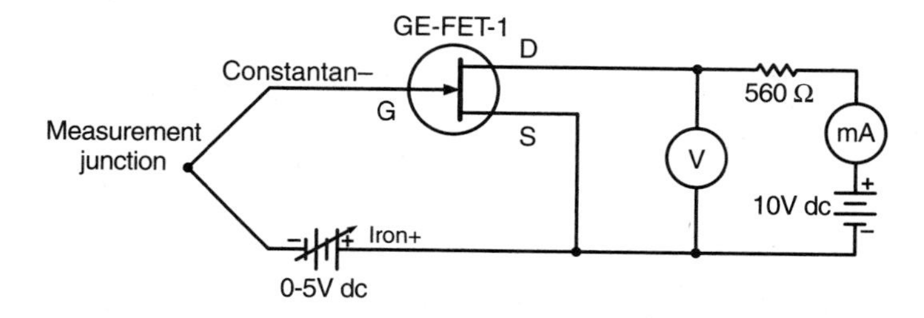

2. Allow the measurement junction of the thermocouple to remain at room temperature. Alter the gate voltage of the FET to equal those listed in **Figure 7-12**. Record the source drain current and voltage for each FET gate voltage value.
3. Adjust the gate voltage to 0.2V.
4. Grasp the measurement junction of the thermocouple between your thumb and forefinger. Describe how this action affects the source drain current and voltage of the FET as compared to the data gathered in Step 2 when the gate voltage was 0.2V.

____

____

Figure 7-12. Thermocouple-controlled FET circuit data.

Gate voltage (V)	Source-drain current (mA)	Source-drain voltage (V)
0.2		
0.4		
0.6		
0.8		
1.0		
1.5		
2.0		

5. Connect the 660 W heat cone to 120V ac and allow it to warm up for about 3 minutes.
6. Position the measurement junction of the thermocouple inside the heat cone for a period of 3 minutes and record the source drain current and voltage of the FET.

   I_{sp} = ______________; V_{sp} = ______________
7. How does the values of source drain current and voltage recorded in Step 6 compare with the data gathered in Step 2 when the FET gate voltage was 0.2V dc?

   ____

   ____
8. Disconnect the heat cone from the 120V ac power supply. What effect does the removal of the heat cone have on the source drain current and voltage?

   ____

   ____
9. Construct the circuit illustrated in **Figure 7-13**.

Figure 7-13. Type J thermocouple transistor circuit.

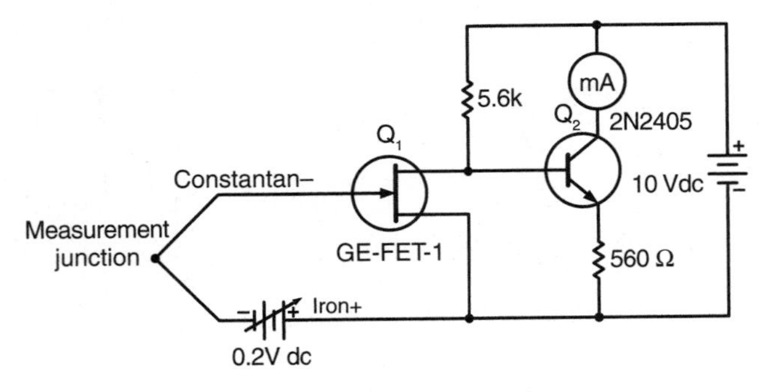

10. Record the collector current of Q_2 as displayed by the digital multimeter with the measurement junction of the thermocouple at room temperature.

    I_c = ______________
11. Grasp the measurement junction of the thermocouple between your thumb and forefinger. Describe how this action affects the collector current of Q_2 as displayed on the digital multimeter.

    ____

    ____

12. Place the measurement junction of the thermocouple inside the cone of the *cool* 660 W heater. Connect the 120V ac to the heat cone and allow it to warm up for 5 minutes.

13. Record the collector current of Q_2 as displayed on the digital multimeter after the heat cone has warmed up.

    I_c = ______________

14. How does the current recorded in Step 10 compare with the current recorded in Step 13?

    ____

    ____

15. How do you explain the difference?

    ____

    ____

16. If the digital multimeter used in the circuit shown in Step 9 were calibrated in degrees Fahrenheit, how could the circuit be used to measure temperature?

    ____

    ____

17. Disassemble and return all components to the storage area.

Analysis:

1. Why is it sometimes necessary to use an amplifier when a thermocouple is used to measure temperature?

   ____

   ____

2. What was the readout device used in the circuit in Step 9?

   ____

3. What would determine the maximum temperature that could be measured by the circuit in Step 9?

   ____

   ____

4. How did the action of the circuit used in Step 9 differ from that used in Step 2?

   ____

   ____

5. Explain how the circuit in Step 2 could be used to measure temperature.

   ____

Activity 9-1—Digital Logic Functions

Name ______________________ Date ______________ Score ____________

Objectives:

Control of a digital system is achieved by a variety of logic gates. Each gate has a particular output response to a combination of input signals. Binary data is represented by 1s and 0s. Multiple-input gates of the AND and OR variety and a single input-output NOT gate represent the three primary logic functions of a digital system. An understanding of these logic functions is essential when analyzing the operation of robotic systems.

In this activity, each basic logic gate will be constructed with a single integrated circuit. Truth tables will then be developed to show the relationship between the input alternatives and the corresponding output. Through this activity, you will gain experience in using the basic logic gates and see how the logic functions are achieved electronically.

Equipment and Materials:

- SN7408 IC.
- SN7411 IC.
- SN7432 IC.
- SN7404 IC.
- 5V dc power supply.
- 470 Ω, 1/8 W resistors (4).
- Light-emitting diodes (4).
- SPDT toggle switches (3).
- SPST toggle switch.
- IC breadboard construction unit.
- Multimeter.

Procedure:

Part A: Gate Testing with the SN7408 and SN7411

1. Connect the logic gate test circuit of **Figure 9-1**.
2. Before closing the circuit switch, turn on the power supply and adjust it to 5V.
3. Turn on the circuit switch and test the IC gate according to the alternatives listed in the truth table of **Figure 9-2**.

Figure 9-1. SN7408 IC test circuit and truth table.

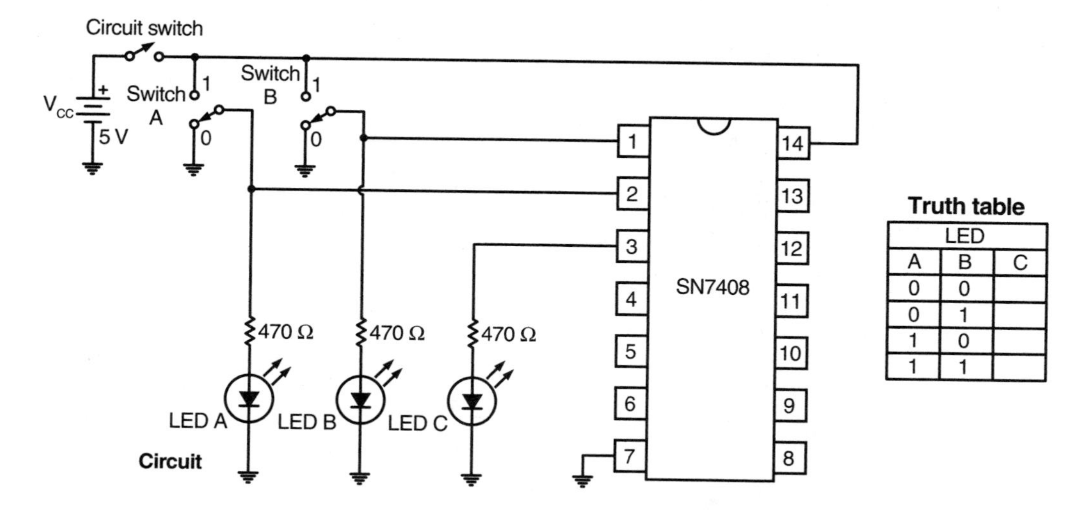

LED		
A	B	C
0	0	
0	1	
1	0	
1	1	

Figure 9-2. SN7411 IC test circuit and truth table.

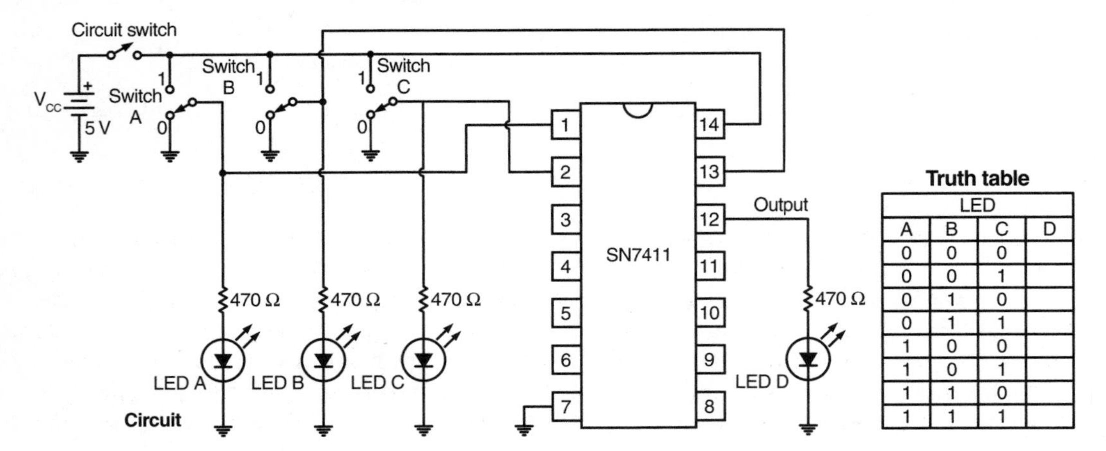

LED			
A	B	C	D
0	0	0	
0	0	1	
0	1	0	
0	1	1	
1	0	0	
1	0	1	
1	1	0	
1	1	1	

4. Prepare a multimeter to measure voltage and determine the corresponding input and output voltages corresponding to 1 and 0.
5. There are three other gates included in this IC chip. Pins 4 and 5 are the input and pin 6 is the output of gate 2. Pins 9 and 10 are the input and pin 8 is the output of gate 3. Pins 13 and 12 are the input and pin 11 is the output of gate 4. Test these gates to see if they are working properly.
6. Make a gate symbol drawing on the blank IC dual inline package layout of **Figure 9-2** for the SN7408.
7. Open the circuit switch and remove the SN7408. In its place connect an SN7411. Be certain that pin 14 is connected to the +5V v_{cc} source and pin 7 is connected to ground. Connect the remainder of the circuit indicated in **Figure 9-2**.
8. Close the circuit switch and test the SN7411. Complete the truth table for the listed input alternatives.
9. Open the circuit switch and disconnect the IC.

Part B: Gate Testing with the SN7432

1. Construct the logic gate test circuit shown in **Figure 9-3**.
2. Before closing the circuit switch, turn on the power supply and adjust it to 5V.
3. Turn on the circuit switch and test the IC gate according to those alternatives listed in **Figure 9-3**.
4. What logic gate function is achieved by this gate?

____

5. There are three other gates included in this IC chip: pins 4 and 5 are inputs; pin 6 the output; pins 9 and 10 are inputs; pin 8 is an output; and pins 12 and 13 are inputs; pin 11 the output. Test these gates to see if they are working properly.

Figure 9-3. SN7432 IC test circuit and truth table.

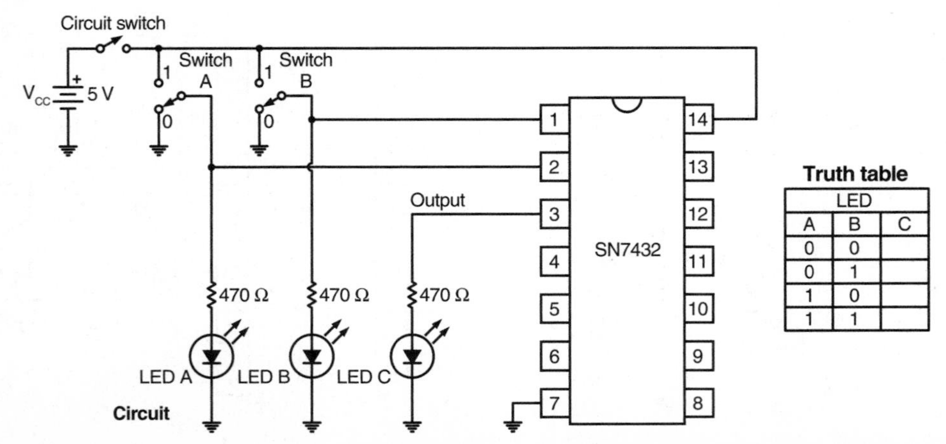

LED		
A	B	C
0	0	
0	1	
1	0	
1	1	

Name ______________________________

6. Make a gate symbol drawing on the blank IC dual inline package layout of **Figure 9-4** showing the input-output connections for the SN7432.

Figure 9-4. SN7432 IC pin connections.

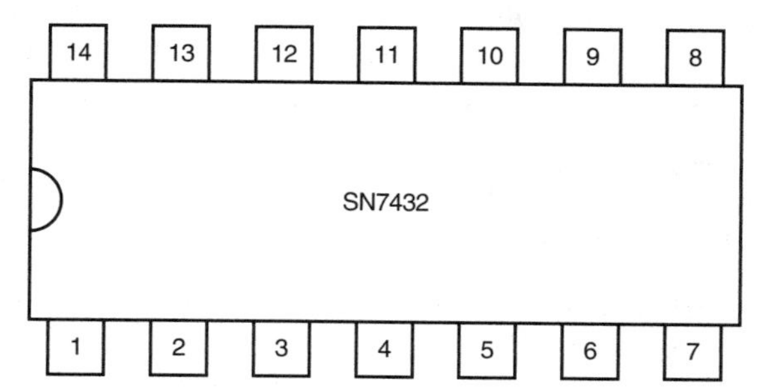

7. Open the circuit switch and disconnect the circuit.

Part C: Gate Testing with the SN7404

1. Connect the logic gate test circuit of **Figure 9-5**.

Figure 9-5. SN7404 IC test circuit and truth table.

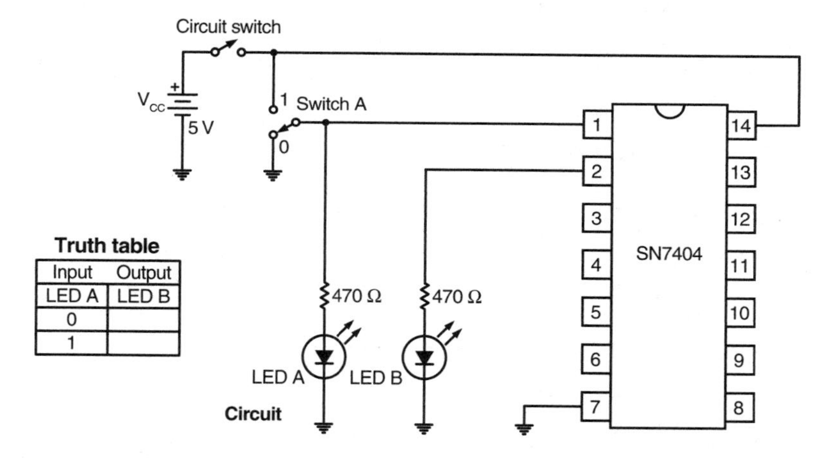

Truth table

Input	Output
LED A	LED B
0	
1	

2. Before closing the circuit switch, turn on the power supply and adjust it to 5V dc.
3. Then close the circuit switch and test the IC gate according to the alternate input conditions listed in the truth table of **Figure 9-5**. Record the respective output for each of the inputs.
4. What logic function is achieved by this gate?

____

5. Test the remaining five gates at pins 3-4, 5-6, 9-8, 11-10, and 13-12. (The pin number designation listed first is the input, with the last number indicating the output.)
6. Make a gate symbol drawing on the blank dual inline package layout of **Figure 9-6** showing the input-output connections for each gate of the SN7404.
7. Open the circuit switch and connect the output of gate 1 to the input of gate 2. Connect the output indicating LED to pin 4.
8. Close the circuit switch and test the logic circuit. What function does this indicate?

____

____

9. Open the circuit switch and disconnect the circuit. Return all parts to the storage area.

Figure 9-6. SN7408 IC pin connections.

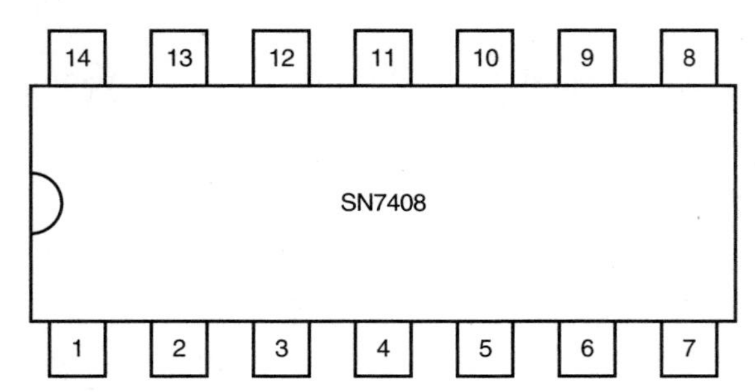

Analysis:

1. What logic functions are achieved by the gates studied in Part A?

2. What mathematical functions are achieved by the gates studied in Part A?

3. What are the symbolic representations of the gates studied in Part A?

4. Prepare a statement that describes the operation of the gates studied in Part A.

5. What mathematical function is achieved by the gate studied in Part B?

6. What are the two logic symbols that are used to represent the circuit constructed in Part B of this activity?

7. Prepare a statement that describes the operation of the gate studied in Part B.

8. What mathematical function is achieved by the logic gate studied in Part C?

9. What is meant by the terms "negation" and "double negation?"

10. What would be the resulting output of a 1 applied to three of the gates studied in Part C of this activity connected in series?

Activity 9-2—Combination Logic Gates

Name ______________________________ Date ______________ Score __________

Objectives:

When either an AND gate or an OR gate is connected to a NOT gate, two additional logic functions are achieved. A NOT-AND, or NAND, is one type of combination logic gate. The NOT-OR, or NOR, gate is representative of the second type of combinational logic gate achieved.

NOR, NAND, and NOT gates are often considered to be universal building blocks in digital systems. With these gates, it is possible to build four logic gates plus the original function. A person working with digital systems should be very familiar with the universal building block principle of combinational logic gates.

In this activity, you will investigate the NAND, NOR, and NOT gates in combinational logic gate construction operations. Only gates that have an inverting capability can be used in this building block technique.

Equipment and Materials:

- SN7400 IC.
- SN7402 IC.
- SN7404 IC.
- SPST toggle switch.
- SPDT toggle switches (2).
- 470 Ω, 1/8 W resistors (3).
- Light-emitting diodes (3).
- 0–5V dc, 1 A power supply.
- Multimeter.
- IC circuit construction board.

Procedure:

Part A: NAND Logic

1. Using the SN7400 quad NAND of **Figure 9-7**, build the IC test circuit and check gate 1.

Figure 9-7. SN7400 IC test circuit and truth table.

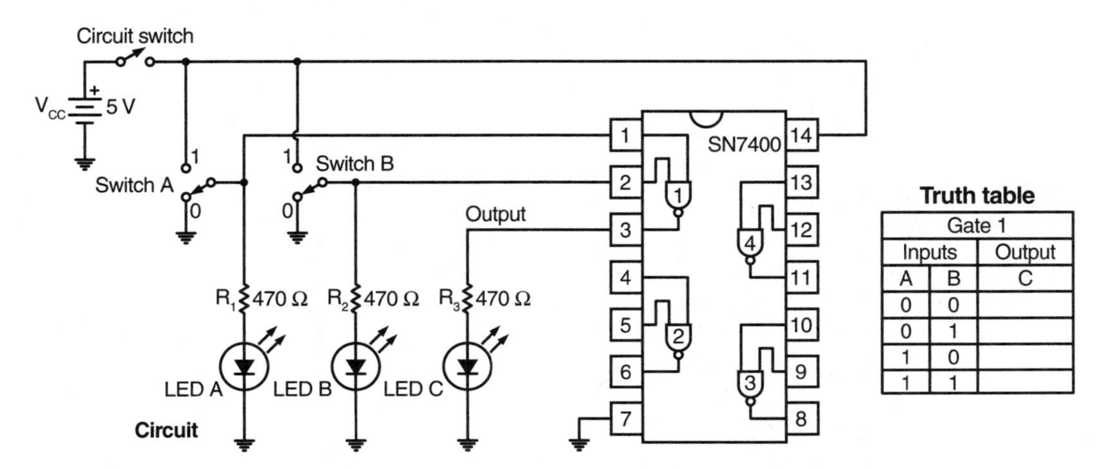

Gate 1		
Inputs		Output
A	B	C
0	0	
0	1	
1	0	
1	1	

2. Record the 1s and 0s of the output in the truth table for each of the input alternatives.
3. With a multimeter, measure and record voltage values represented by the 1s and 0s of this gate.

   A 1 = ______________ V, while a 0 = ______________ V.

4. Test the other three gates to make certain they are functioning properly.
5. To build a NOT gate, remove the B circuit from pin 2 and connect pins 1 and 2 together. Use switch A as the input and test the output of the gate.
6. Connect NAND gate 1 as instructed in procedure Steps 1 and 3. Connect the output of gate 1 to pins 4 and 5 of NAND 2. Connect the output LED to pin 6. What function does this combination logic gate achieve?

____

____

7. Test the circuit to verify your theory. Did it work as you predicted?

____

8. Connect the combinational logic circuit of **Figure 9-8**. Complete the truth table showing the outputs at the indicated points.

Figure 9-8. Combination NAND gate circuit and table.

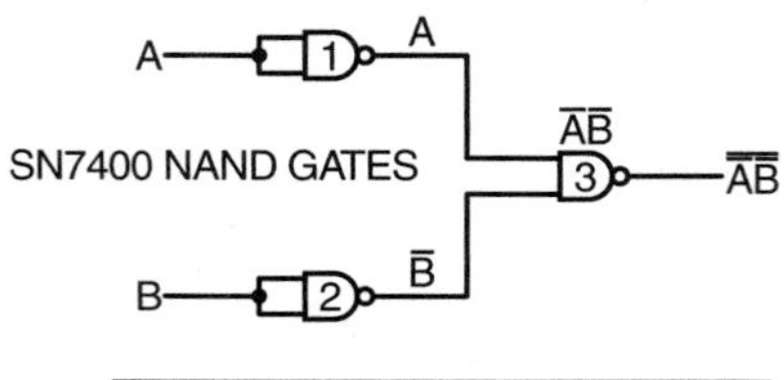

Inputs		Expression			Output
A	B	$\overline{A}$	$\overline{B}$	$\overline{A}\,\overline{B}$	$\overline{\overline{A}\,\overline{B}}$
0	0				
0	1				
1	0				
1	1				

9. According to the truth table, what gate function is achieved by this combination logic gate? Test the gate to verify your prediction. Did it perform the function you expected?

____

Part B: NOR Logic

1. Using an SN7402 quad NOR gate, **Figure 9-9**, build the IC test circuit and check gate 1.
2. Record the 1 and 0 outputs for the input alternatives listed in the truth table in **Figure 9-9**.

Figure 9-9. SN7402 IC test circuit and truth table.

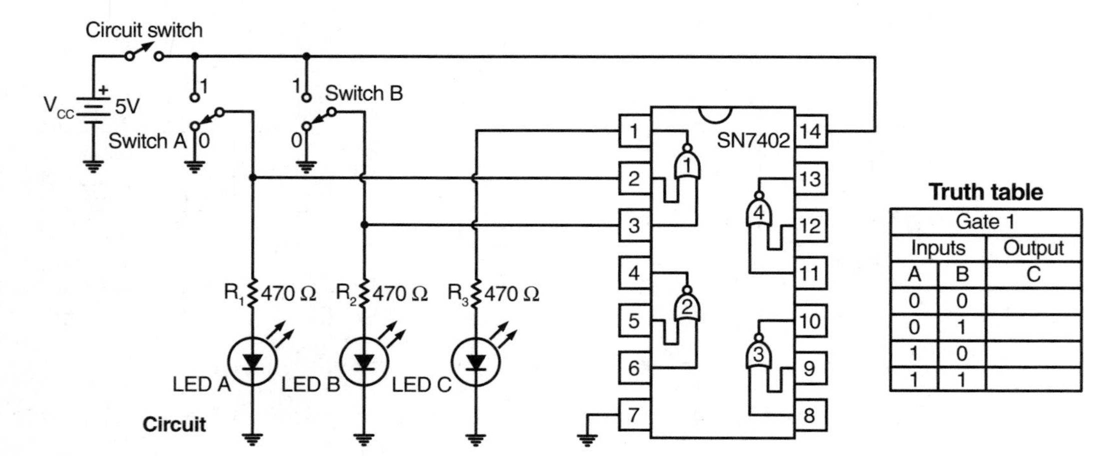

Truth table

Gate 1		
Inputs		Output
A	B	C
0	0	
0	1	
1	0	
1	1	

Name ____

3. With a multimeter, measure the voltage values represented by the 1s and 0s of this gate.
4. Test the other three gates to make certain they are functioning properly.
5. To build a NOT gate, disconnect the lead to pin 3 and then connect pin 2 to pin 3. Use switch A as the input and test the output of the gate.
6. Connect NOR gate 1 as instructed in Steps 1 and 2. Connect the output of NOR 1 to the input of NOR 2, with NOR 2 connected as a NOT gate. Connect the output LED to pin 4. What combination logic is achieved by this configuration?

   ____

   ____

7. Test the circuit to verify your theory. Did it work as you predicted?

   ____

8. Connect the combinational logic circuit of **Figure 9-10** using the SN7402.

Figure 9-10. Combination NOR gate circuit and table.

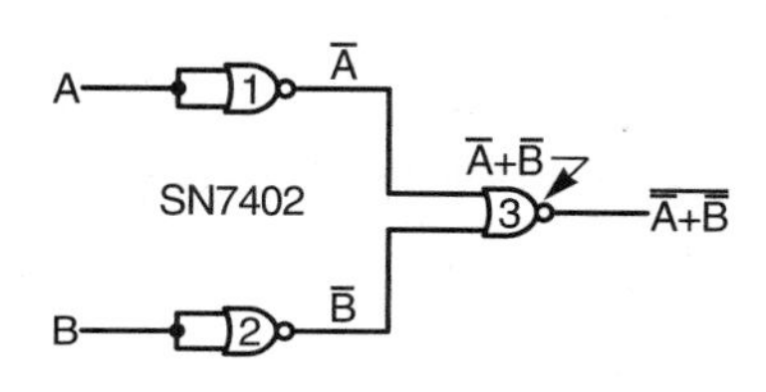

Inputs		Expression			Output
A	B	A	B	$\overline{A}+\overline{B}$	$\overline{\overline{A}+\overline{B}}$
0	0				
0	1				
1	0				
1	1				

9. Complete the truth table in **Figure 9-10** for outputs at the designated points. Then test the combinational logic circuit. Does it do what you predicted by the truth table?

   ____

10. Open the circuit power switch and connect NOR gate 4 as a NOT gate. Attach the output in **Figure 9-10** to the input of NOR 4. Connect the LED to the output of NOR 4. Invert the last stage of the truth table of **Figure 9-10** and predict the type of gate achieved by your modified circuit.
11. Close the circuit switch and test your prediction. Was it correct?

    ____

12. Open the circuit switch and disconnect the circuit.

Part C: NOT Logic

1. Using the SN7404 hex-inverter IC of **Figure 9-11**, build the IC test circuit and check gate 1.
2. Turn on the circuit switch and record the 1s and 0s of the output in the truth table of **Figure 9-11** for the input alternatives.
3. With a multimeter, measure the voltage values of a representative 1 and 0 of this gate.

   A 1 = ____________________ V, while a 0 = ____________________ V.

4. Test the other five gates to verify that they are operating properly.
5. Open the circuit switch and combine gates 1 and 2 as indicated in **Figure 9-12**.

Figure 9-11. SN7404 IC test circuit and truth table.

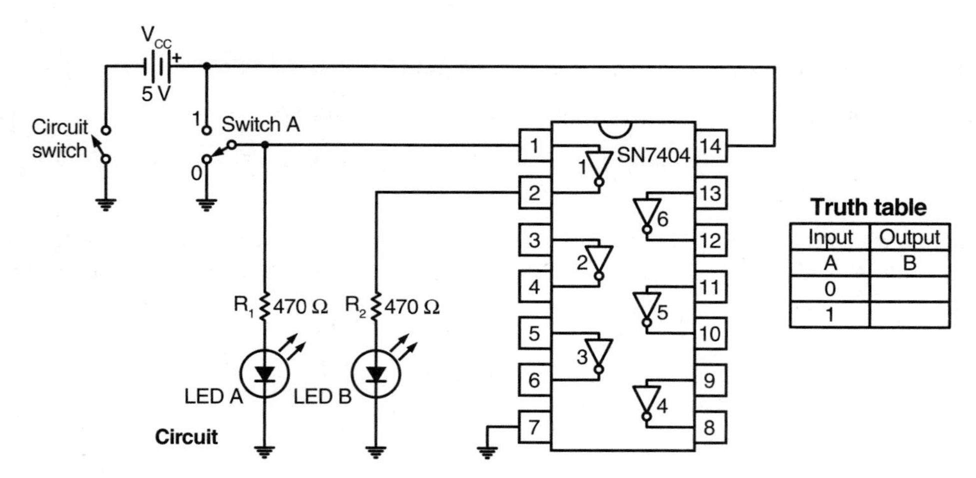

Truth table

Input	Output
A	B
0	
1	

6. Close the circuit switch and test the combination gate. Record the 1 and 0 outputs for each of the input alternatives of the truth table in **Figure 9-12**. What logic function does this achieve?

____

____

7. Open the circuit switch and connect the output of the circuit in **Figure 9-12** to the input of gate 3. Connect the LED-resistor to pin 6 and test the circuit. What logic function does it achieve?

____

____

Figure 9-12. Combination NOT gate circuit and table.

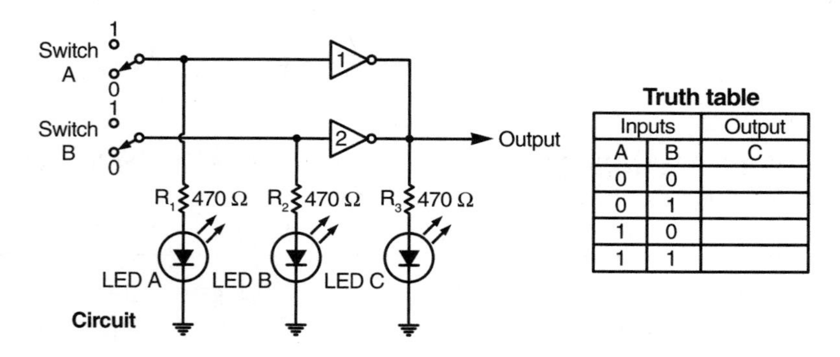

Truth table

Inputs		Output
A	B	C
0	0	
0	1	
1	0	
1	1	

8. Open the circuit switch and combine gates of the inverter to form the circuit of **Figure 9-13**.

Figure 9-13. Cascaded double-inverter circuit.

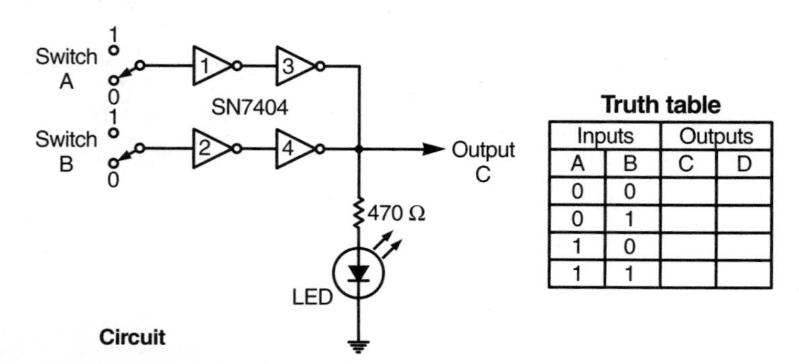

Truth table

Inputs		Outputs	
A	B	C	D
0	0		
0	1		
1	0		
1	1		

Name ____

9. Close the circuit switch and test the circuit. Record the 1 and 0 outputs of C in the truth table of **Figure 9-13** for the given input alternatives. What gate function does this achieve?

____

____

10. Open the circuit switch and add the output of **Figure 9-13** to the input of gate 5. Move the LED to the output of gate 5. What gate function do you predict the output will demonstrate?

____

____

11. Close the circuit switch and test the circuit. Record your findings at the D output of the truth table in **Figure 9-13**. Was your prediction correct?

____

12. Open the circuit switch and disconnect the components. Return all parts to the storage area.

Analysis:

1. In Part A, complete **Figure 9-14** showing the outputs for each logic gate listed.

Figure 9-14. General truth table.

Inputs		Outputs			
A	B	NAND	AND	OR	NOR
0	0				
0	1				
1	0				
1	1				

2. Make a sketch of the five logic circuit combinations accomplished with the SN7402. Show the actual logic gate combinations used.

3. In any of the combined inverter gate combinations, why is the 0 considered a predominant factor?

____

____

4. Make a sketch of the five logic functions achieved by the SN7404.

Activity 10-1—Electromagnetic Relays

Name ______________________________ Date ______________ Score ____________

Objectives:

Relays are electromagnetic switches and are excellent examples of how a magnetic field attracts a magnetic material. These devices contain a coil that creates an electromagnetic field, an armature, which is constructed of a magnetic material attracted by the coil, and a number of contacts or switches that open or close when the magnetic field attracts the armature.

In this activity, you will study the electromagnetic characteristics of a relay. Relays are a popular type of control device that might be used with robotic systems.

Equipment and Materials:

- Multicontact relay.
- 6V lamp with socket.
- Variable dc power supply.
- Resistor—1 kΩ.
- 6V battery.
- Connecting wires.
- Multimeter.

Procedure:

1. Prepare the multimeter to measure resistance. Measure and record the resistance of the relay coil:

   ______________ Ω

2. Using the multimeter, determine how many normally open and normally closed contacts are used with your relay.

   Number of normally open contacts = ______________

   Number of normally closed contacts = ______________

3. Construct the circuit illustrated in **Figure 10-1**. Be sure that the variable dc power supply is adjusted to zero. The multimeter should be adjusted to measure dc current on the highest range.

4. Slowly adjust the variable dc power supply from zero until the 6V lamp is turned on. Record the current measured by the multimeter when the relay is energized. This is the pickup current:

   Pickup current = ______________ mA

Figure 10-1. Circuit using electromagnetic relay.

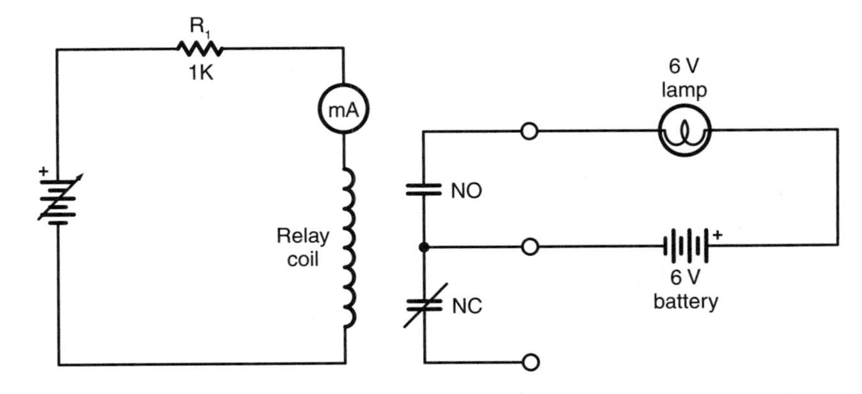

5. Slowly adjust the variable dc power supply toward zero until the 6V lamp is turned off. Record the current measured by the multimeter when the relay de-energized. This is the dropout current:

   Dropout current = ____________________ mA

6. Turn the variable power supply off.
7. Alter the circuit to cause it to be like that in **Figure 10-2**.
8. You will notice that the only difference in the two circuits is the type of contacts used. In Step 3, the normally open contacts were used. In this procedure, the normally closed contacts are used, causing the lamp to remain on until the relay is energized.

Figure 10-2. Circuit from **Figure 10-1** with relay contacts reversed.

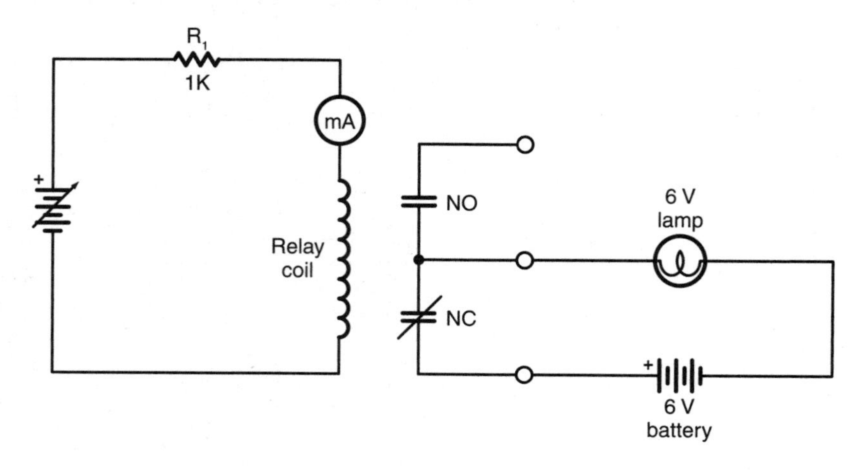

9. Adjust the variable dc power supply and record the pickup and dropout currents as you did in Steps 4 and 5.

   Pickup I = ____________________ mA

   Dropout I = ____________________ mA

10. How do the currents recorded in Step 9 compare with the current recorded in Steps 4 and 5?

    ____

11. How did the action of the 6V lamp in Steps 4 and 5 compare with the action of the lamp in Step 9?

    ____

Analysis:

1. What are normally open contacts?

   ____

2. What are normally closed contacts?

   ____

3. What is meant by the term pickup current?

   ____

4. What is meant by the term dropout current?

   ____

5. Using Ohm's law, compute the voltage across the relay coil when the relay is energized (see Steps 1 and 4):

   $V = I \times R =$ ____________________ V

Activity 11-1—Soldering and Terminal Connections

Name ______________________________ Date ______________ Score ___________

Objectives:

In this activity, you will demonstrate your skill in three important exercises: (1) soldering, (2) desoldering, and (3) terminal connection. An understanding of the proper soldering and terminal connection techniques is needed by technicians who design, service, or repair industrial equipment, including robots.

Equipment and Materials:

- Soldering iron—25 W–50 W.
- Desoldering tool.
- Solder.
- No. 22 insulated wire—3″ length.
- Printed circuit board.
- Component mounting strip.
- Resistor (any value).
- Wire stripper.
- Side-cutting pliers.
- Needle-nose pliers.
- "Solderless" connector and wire.
- Crimping tool.

Safety:

Exercise caution while soldering. Use protective eyewear and appropriate clothing during soldering operations.

Procedure:

Section A. Soldering Iron Preparation and Component Mounting

1. Allow the soldering iron to heat to its operating temperature.
2. Apply a small amount of solder to coat (this is called 'tinning') the tip of the iron.
3. Obtain a printed circuit (pc) board and a 3″ length of No. 22 insulated wire or a resistor.
4. Place the wire (or resistor) near the holes in the pc board where you intend to solder it to determine the length required.
5. Strip 3/8″ to 1/2″ of insulation from one end of the wire (if used) with the wire stripper.
6. Cut off any excess wire and strip the insulation off the other end of the wire. The wire should lie flat on the pc board when soldered. If a resistor is used, it should lie flat also.
7. Using the proper soldering techniques, solder both ends of the wire (or resistor) to the pc board terminals.

Instructor's Approval: ______________

Section B. Printed Circuit Board Desoldering

1. Obtain a desoldering tool.
2. Allow the desoldering tool to heat to its operating temperature.
3. Using the proper desoldering techniques, remove the piece of wire from the pc board without damaging the conductive strips on the pc board.

Instructor's Approval: ____________________

Section C. Terminal Connection Installation

1. Obtain a "solderless" terminal connector crimping tool and a short length of insulated wire from your instructor.
2. Strip enough insulation from the wire so the terminal connector fits properly.
3. Use the crimping tool to fasten the terminal connector to the wire.

Instructor's Approval: ____________________

Activity 11-2—Troubleshooting and Testing

Name ______________________________ Date ______________ Score __________

The lab activities included in this book provide an opportunity to practice troubleshooting and testing electronic circuits, devices, and systems. Troubleshooting and testing procedures may be accomplished while performing the lab activities. Emphasis is placed on understanding circuit operation, safety, and the proper use of test equipment. If the technician understands how the circuit, device, or system functions and knows how to use test equipment, troubleshooting and testing are relatively easy to accomplish. This is true for the simplest type of circuit or for more complex systems.

Objectives:

Upon completion of this activity, you should be able to outline basic troubleshooting procedures, test devices and circuits using the correct procedure, and find the defective part or circuit by using a "common sense" approach.

Troubleshooting is a systematic process by which malfunctions are traced and repaired. A technician should be able to systematically locate and repair many malfunctions. For effective troubleshooting, you should have a practical understanding of the type of system that is being repaired. With this basic knowledge of operation and some test equipment, you should be able to perform many repair jobs.

Electrical motors are commonly used with robotic systems. Some of the following instruments may be used to make various tests on the motor: (1) tachometer—speed test; (2) megger—insulation test; (3) multimeter—resistance, current, and voltage test; (4) growler—wound armature test; and (5) motor dynamometer analysis unit—speed vs torque test. You should check with the instructor on the use of some of this equipment as your troubleshooting progresses.

After you complete this activity, you will understand how to systematically test and repair a motor of any type. Many parts of the procedure listed may not be applicable to your particular troubleshooting job. However, try to follow most of the listed steps to complete the analysis although it may not be feasible to complete the entire activity.

Equipment and Materials:

- Motor that does not function properly.
- Multimeter.
- Growler (optional).
- Megohmmeter (optional).
- Dynamometer (optional).
- Tachometer (optional).

Procedure:

1. Examine the external features of the motor you are to troubleshoot.
2. Find the motor nameplate and record the information called for below. If not available, indicate so in the blank.

   Manufacturing Co. ______________________________

   Motor Type ______________________________

   Identification Number ______________________________

   Model Number ______________________________

   Frame Type ______________________________

   Number of Phases (AC) ______________________________

Horsepower ______________________________

Cycles (AC) ______________________________

Speed (rpm) ______________________________

Voltage Rating ______________________________

Current Rating (Amperes) ______________________________

Thermal Protection ______________________________

Temperature Rating ______________________________

Time Rating ______________________________

Other Information ______________________________

____

____

____

____

3. Determine the type of motor you are working with, such as ac induction, dc stepping, shaded-pole, ac servo. Consult with your instructor if you cannot determine the specific type. What is the specific motor type?

____

4. In the following space, draw a schematic diagram of the motor. This will provide a general reference for later tests.

5. After the proper type of power source and operating voltage has been determined, apply the correct voltage to the motor for a very brief period of time while attempting to detect the specific malfunction.
6. Describe any malfunctions that seem to exist in your motor.

____

____

____

____

Name ____

7. If available, consult a symptom chart in a motor operation book, manufacturer's data or other reference. List some possible causes of the malfunction.

____

____

____

____

8. Determine and list some possible corrections for the malfunction.

____

____

____

____

9. If the motor must be disassembled, remove the end plates carefully. Be particularly careful with the connecting wires and bearings. Remove the rotor assembly at this time. Check the bearings.
10. If connecting wires must be removed, use tape, or some other means to label the proper terminal connections.
11. Inspect all wiring connections, splices, and other mechanical connections for possible opens or shorts.
12. Inspect the stator portion of the motor for excessive dirt, oil, or other damage.
13. Check the rotor for visible damage.
14. If the rotor is the squirrel-cage type, it may be in need of cleaning with fine sandpaper.
15. If a wound rotor is used, check the commutator for high mica insulation, roughness, irregular wear, or other damage. You may wish to dress the commutator on a metal lathe, clean it with sandpaper, or remove high mica with an undercutting tool.
16. If the motor has brushes, check for excessive wear, oil, grease, chipping, or improper contact.
17. You may need to perform some resistance tests with the multifunction meter to determine the condition of the stator windings. (Example: Ac single-phase induction motor; start winding—20 ohms resistance; run winding—5 ohms resistance.) Of course, those resistances will vary. Record your resistance readings.

    Winding resistances = ____

    ____

18. What can you conclude from the resistance test regarding the condition of your motor?

____

____

____

____

19. You may test the rotor windings for damaged insulation with a megohmmeter. You may test the rotor for shorts, grounds, or open circuits with a test instrument called a 'growler' if available.
20. Perform any other tests you feel may be needed to locate the trouble. Consult with your instructor.
21. Perform the necessary procedures to repair the motor.
22. Consult with your instructor when you feel you have completed the troubleshooting procedure and have repaired the motor malfunction.

Instructor's Approval: ____________________

23. This should complete your troubleshooting job. Return all materials to the storage area.

Analysis:

1. Write a detailed description of the troubleshooting process you have used in this experiment. Explain how you determined the trouble for this particular type of motor. Also, explain how you corrected (or would correct) the malfunction.

Fluid Power Symbols

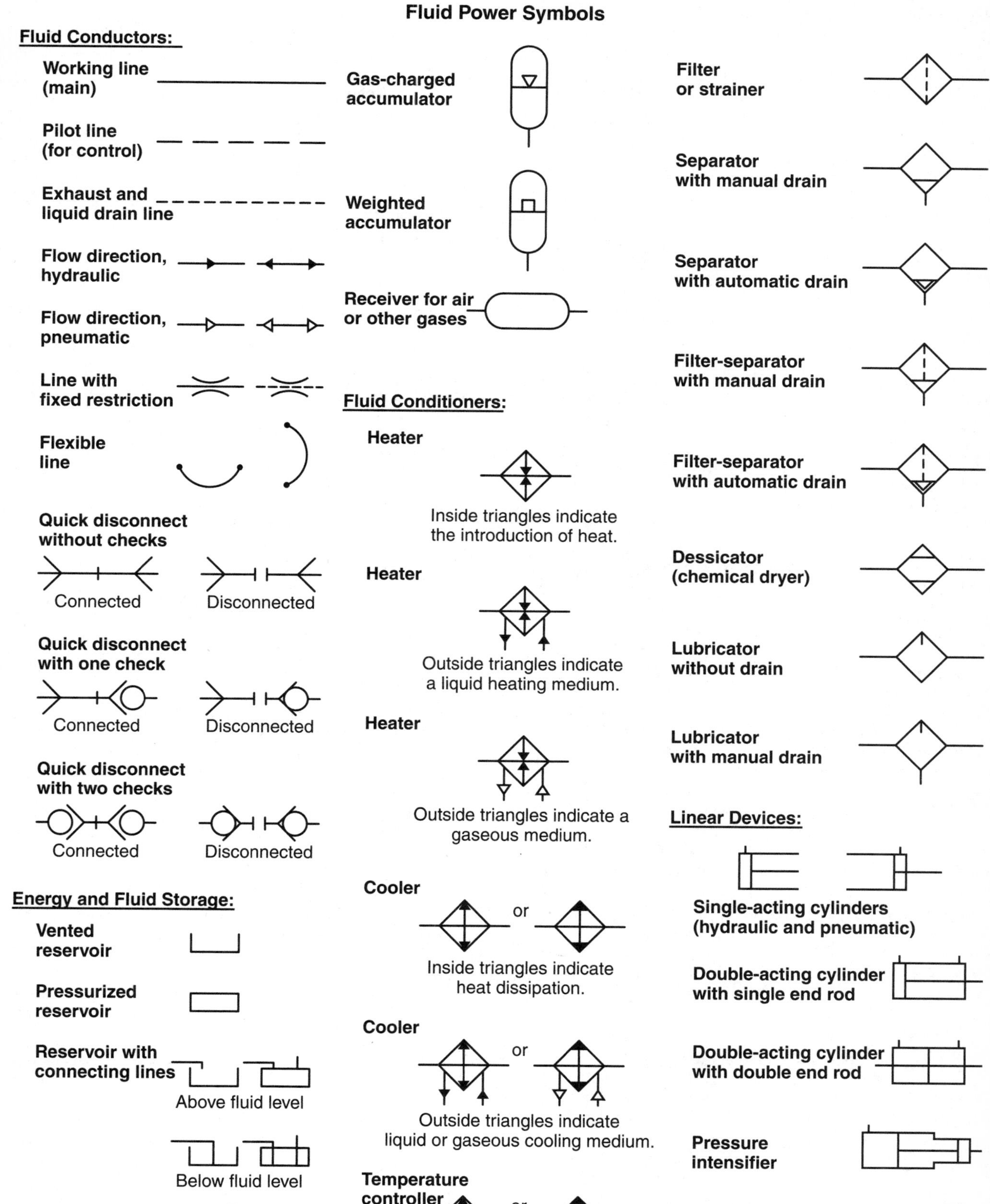

Temperature controller

or

Outside triangles indicate a liquid or gaseous cooling medium.

Spring-loaded accumulator

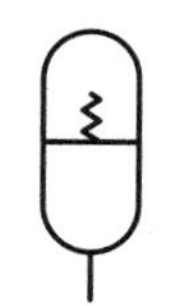

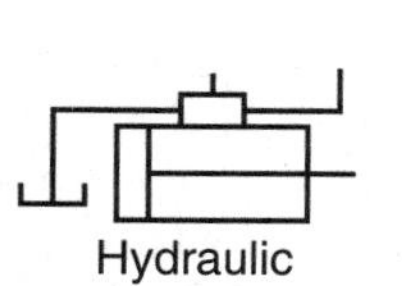

Hydraulic

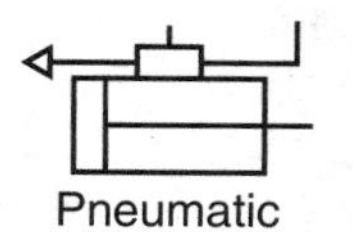

Pneumatic

Servo positioner

Actuators and Controls:

Spring

Manual

Push button

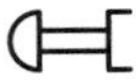

Lever

Pedal or treadle

Mechanical

Detent

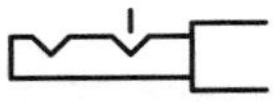

Short line indicates detent in use.

Pressure compensated

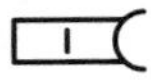

Solenoid (single winding)

Reversing motor

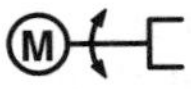

Pilot pressure (remote supply)

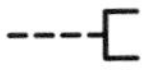

Pilot pressure (internal supply)

Actuation by released pressure

By remote exhaust | By internal return

Pilot controlled, spring centered

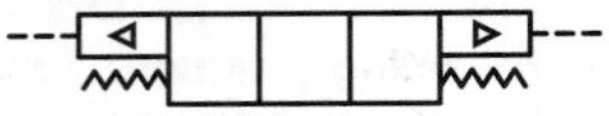

Pilot differential

Solenoid or pilot

External pilot supply | Internal pilot supply and exhaust

Solenoid and pilot

Actuation by thermal change

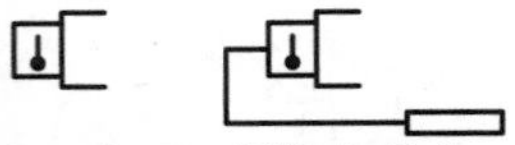

Local sensing | With bulb for remote sensing

Servo

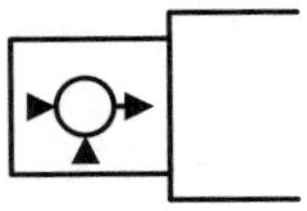

Rotary Devices:

Unidirectional | Bidirectional

Hydraulic pump, fixed displacement

Unidirectional | Bidirectional

Hydraulic pump, variable displacement, noncompensated

Unidirectional | Bidirectional

Hydraulic pump, variable displacement, pressure compensated

Unidirectional | Bidirectional

Hydraulic motor, fixed displacement

Unidirectional | Bidirectional

Hydraulic motor, variable displacement

Operates as pump in one direction, as motor in other direction.

Hydraulic pump–motor

Operates in one direction, as either pump or motor.

Hydraulic pump–motor

Operates in both directions, as either pump or motor.

Hydraulic pump–motor, variable displacement, pressure compensated

Pneumatic pump, fixed displacement — Compressor

Pneumatic pump, fixed displacement — Vacuum pump

Pneumatic motor

Unidirectional | Bidirectional

Oscillator

Hydraulic | Pneumatic

Electric motor

Internal combustion engine

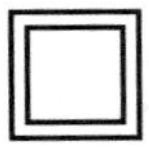

Instruments and Accesories:

Pressure indicating and recording

Temperature indicating and recording

Flow-rate meter

Totalizing meter

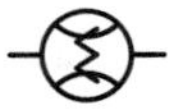

Venturi

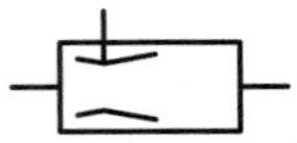

Orifice plate

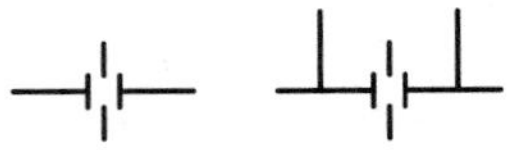

Pitot tube

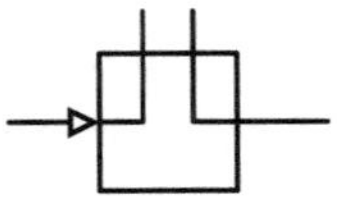

Nozzle

Hydraulic | Pneumatic

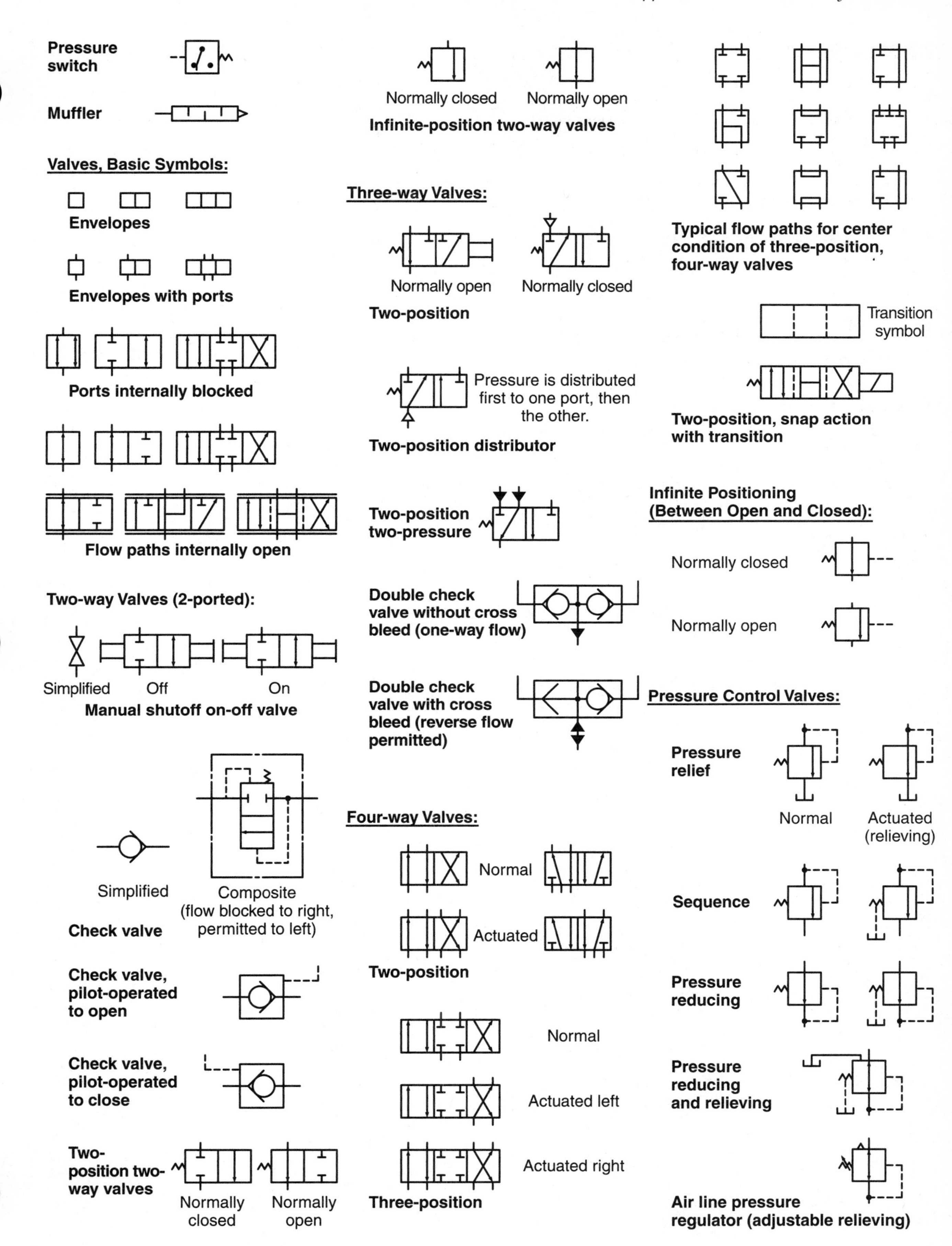
Pressure switch
Muffler
Valves, Basic Symbols:
Envelopes
Envelopes with ports
Ports internally blocked
Flow paths internally open
Two-way Valves (2-ported):
Simplified
Off
On
Manual shutoff on-off valve
Simplified
Composite
(flow blocked to right, permitted to left)
Check valve
Check valve, pilot-operated to open
Check valve, pilot-operated to close
Two-position two-way valves
Normally closed
Normally open
Normally closed
Normally open
Infinite-position two-way valves
Three-way Valves:
Normally open
Normally closed
Two-position
Pressure is distributed first to one port, then the other.
Two-position distributor
Two-position two-pressure
Double check valve without cross bleed (one-way flow)
Double check valve with cross bleed (reverse flow permitted)
Four-way Valves:
Normal
Actuated
Two-position
Normal
Actuated left
Actuated right
Three-position
Typical flow paths for center condition of three-position, four-way valves
Transition symbol
Two-position, snap action with transition
Infinite Positioning (Between Open and Closed):
Normally closed
Normally open
Pressure Control Valves:
Pressure relief
Normal
Actuated (relieving)
Sequence
Pressure reducing
Pressure reducing and relieving
Air line pressure regulator (adjustable relieving)

Infinite Positioning Valves:

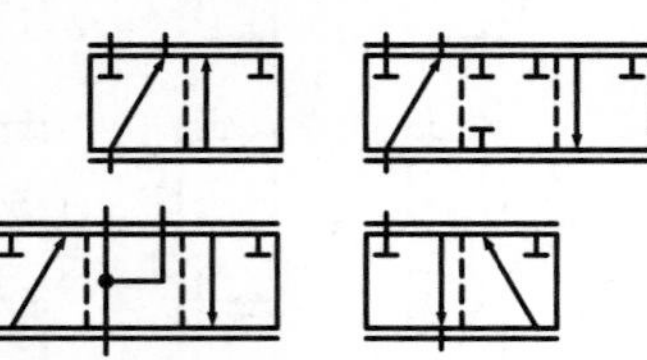

Three-way valves

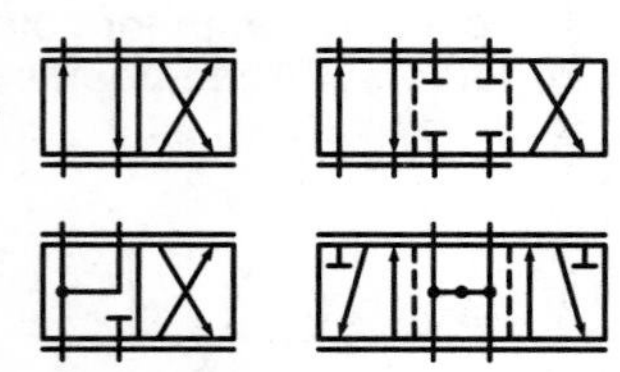

Four-way valves

Flow Control Valves:

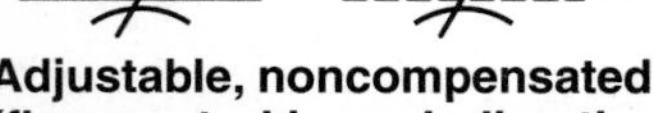

Adjustable, noncompensated (flow control in each direction)

Adjustable with bypass

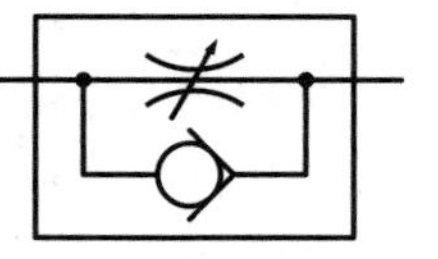

Flow controlled to right, flow to left bypasses control.

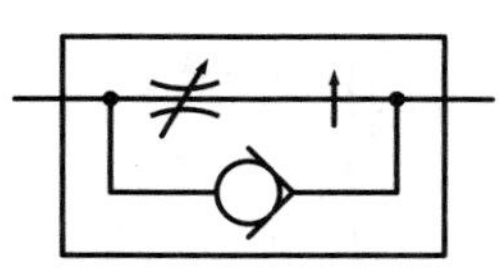

Adjustable and pressure compensated, with bypass

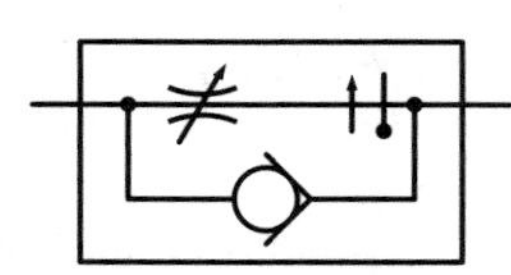

Adjustable temperature and pressure compensated

Air Line Accessories:

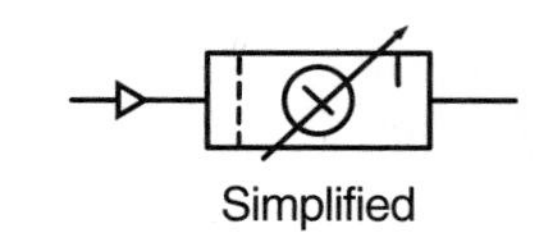

Simplified

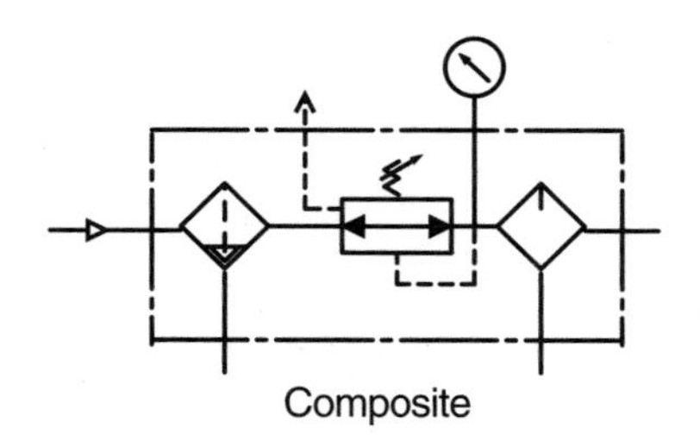

Composite

Filter regulator, and lubricator